Also by Bill Carter

Boom, Bust, Boom:
A Story about Copper, the Metal That Runs the World

Fools Rush In:
A True Story of Love, War, and Redemption

RED SUMMER

The Danger, Madness, and Exaltation of Salmon Fishing in a Remote Alaskan Village

Bill Carter

Scribner
New York London Toronto Sydney New Delhi

SCRIBNER
An Imprint of Simon & Schuster, Inc.
1230 Avenue of the Americas
New York, NY 10020

First Scribner trade paperback edition August 2021

For information about special discounts for bulk purchases, please contact Simon & Schuster Special Sales at 1-866-506-1949 or business@simonandschuster.com.

The Simon & Schuster Speakers Bureau can bring authors to your live event. For more information or to book an event, contact the Simon & Schuster Speakers Bureau at 1-866-248-3049 or visit our website at www.simonspeakers.com.

Designed by Kyoko Watanabe

Manufactured in the United States of America

1 3 5 7 9 10 8 6 4 2

Library of Congress Cataloging-in-Publication Data is available.

ISBN 978-0-7432-9706-6
ISBN 978-0-7432-9707-3 (pbk)
ISBN 978-1-4165-6604-5 (ebook)

Certain names and identifying characteristics have been changed.

Photographs were provided courtesy of the author, except for the photograph on page 123, which appears courtesy of Sharon Hart, and the photograph on page 224, which appears courtesy of Griffin Adams.

For Leigh

CONTENTS

RED SUMMER

"What gives value to travel is fear."

—Albert Camus

PROLOGUE

I LIE MOTIONLESS IN THE DARK, SWEATING THROUGH MY CLOTHES, trying to remember where I am. My eyes are open but I cannot see anything other than the time on the clock, 4:00 A.M. My brain feels slushy, in a lucid dream state no doubt aided by the codeine and shots of whiskey I swallowed before lying down four hours ago. The rain hasn't stopped for days. The leak just beyond my bed keeps up a steady trickle of water, now slowly flooding the hallway. No bigger than a standard prison cell, the room has no windows and two single bed frames, each with sunken mattresses and pillows that reek of mothballs. There is a scent of urine in the air, left over from a wintering porcupine. On the far wall a square piece of plywood covers what used to be a window with a view of the river, nailed shut after a bear tried to enter, long before I ever slept here.

Staring into complete darkness I can feel the rocking motion of the boat. It won't go away. Not until the season is over. As for the sweating, it comes from the dream, one that only happens when I'm in Egegik. In it the river is always the culprit, a faceless murderer. I'm never a tragic victim or heroic figure in my dream, just an alert bystander with hands at the ready grabbing the rails of the boat waiting to help, waiting to be useful. And then it happens. No matter how much I foresee what is coming, my feet get caught and I slip into the net. I go overboard, eaten alive by the net and the fish alike as I sink to the bottom of the river. For a split second I can see the terror-stricken faces of my friends, my crewmates, as they scream and reach into the swirling current to grab me. Looking up I can see them pull on the net with all their might. Yet it's too late and I fall to the bottom of the river, gasping my last breath.

Panting, my body covered in sweat, I turn on a small lamp, burying the dream. It is 40 degrees; my breath has become a steady stream of white vapor. I put on my work uniform. First, a long-sleeved long johns top, followed by a T-shirt over that and a thick hooded sweatshirt. Then I pull on my sweatpants, pulling the socks over the top of the sweatpant legs, helping to keep them in place when I put on the waders.

My breath gives me pause: it's deep, rhythmic, and peaceful. I extend my arms. A tightness. It feels like someone has stuck a dozen needles into the underside of my forearm. Slowly, I flex my fingers and the calluses on my fingertips crack, blood trickles out.

Outside, I look over the muddy brown river, the air thick with mosquitoes. Today will be the peak. I've witnessed this spectacle for four summers now. It is a brilliant event; today the salmon are coming with all the might of their existence, pushing for survival, but yielding to death. Gladly yielding? I don't know, I can't read their eyes, yet. Out beyond the river, the snow-covered volcano stands in the distance, like a lone pyramid in the vast Sahara of tundra. Bear tracks the size of dinner plates line the river's edge, the beast having already slipped away for a nap. One day, when man disappears from the planet, this land, this river, and most likely these fish will still be here. For there is a patience in Egegik, one found only in earth's harshest environments, where the outside world has no impact on what happens day to day. Out here the animals and humans share the wilderness, wary of each other, but neighbors nonetheless. Their movements, like a slow form of tai chi, are in direct contact with the earth's pulse, a connection most of us lost long ago.

Calm and warm, I flex my arms, preparing for what will come. I don't belong here. No one belongs here, except possibly the man walking drunk down the dirt path with the .357 in his waistband, his weapon of choice in case the grizzly wants him this night. Of course the gun is no use. The bear would tear him to shreds. This is a place where nature goes about the business of being violent, without apology. Everything up here experiences a harsh death, humans included. No one who stays here ever ends up in a hospice. No one drinks green tea and reads self-help books. The closest thing to an espresso is yesterday's coffee warmed up and served in a plastic cup, likely scavenged from the river. This is a land of extremes and those who keep returning follow the silent restriction that acts as the only social law: *Do the work or leave.*

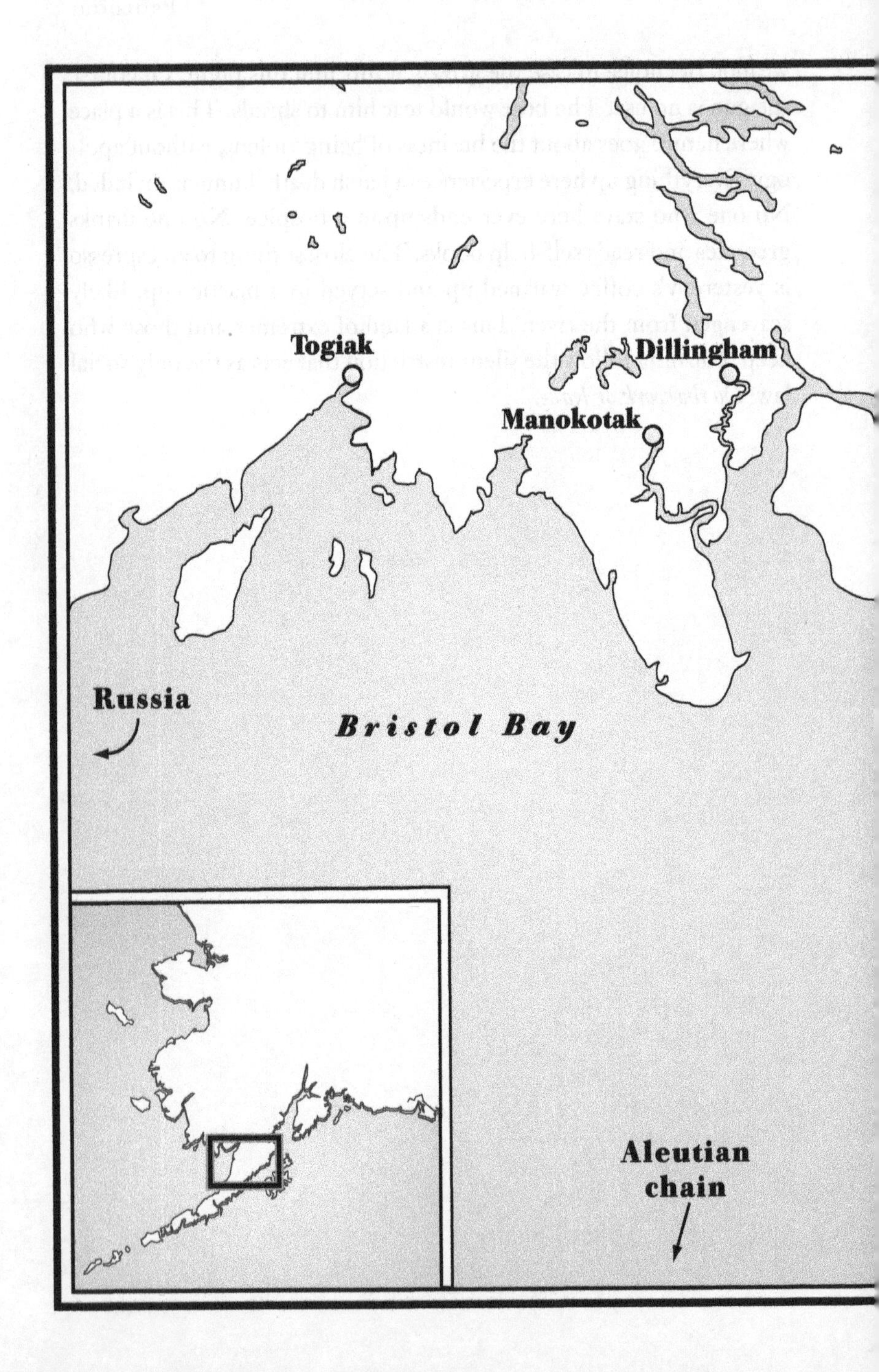
Togiak
Dillingham
Manokotak
Russia
Bristol Bay
Aleutian
chain

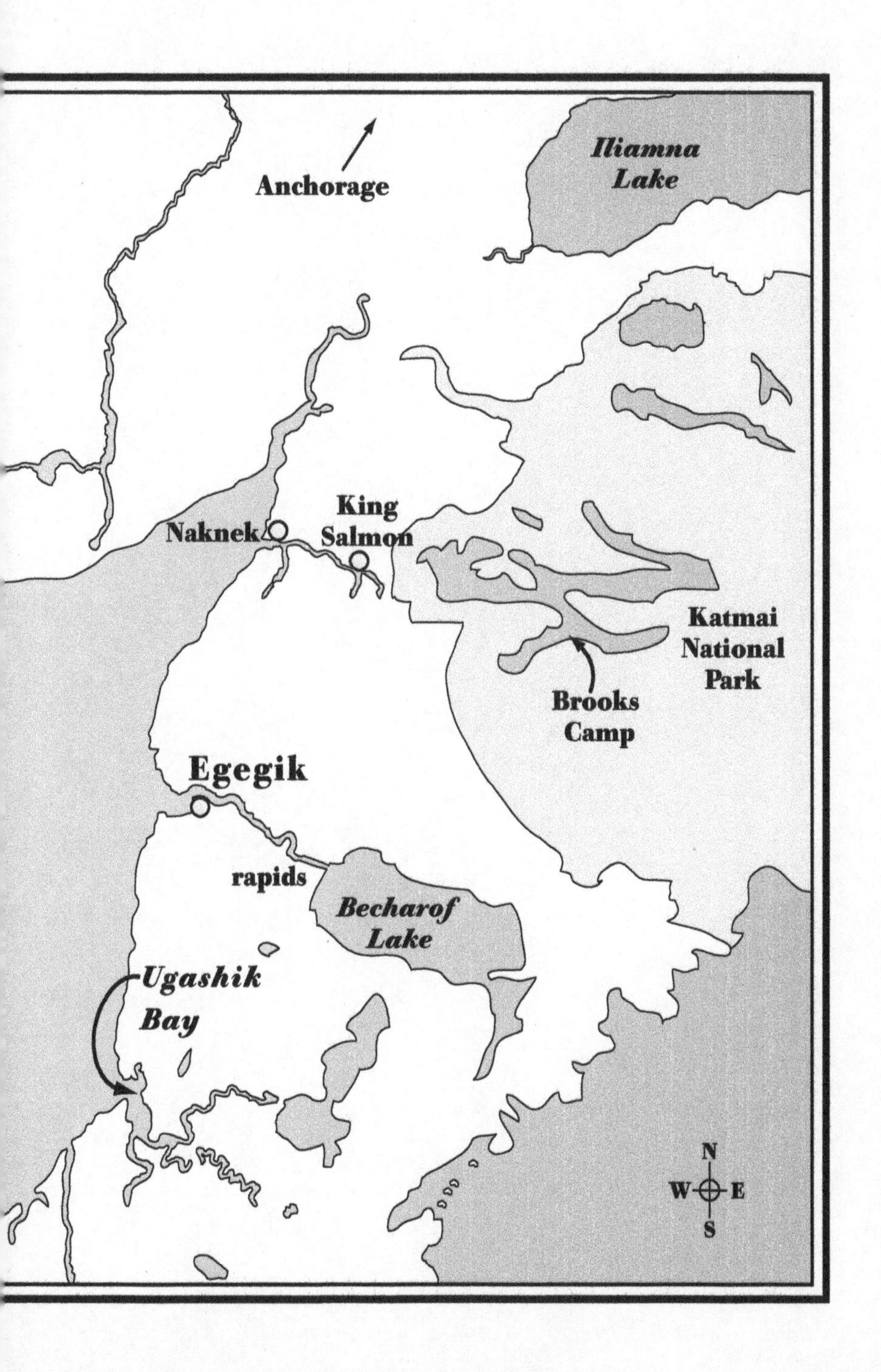

Anchorage
Iliamna Lake
King Salmon
Naknek
Katmai National Park
Brooks Camp
Egegik
rapids
Becharof Lake
Ugashik Bay
N
W
E
S

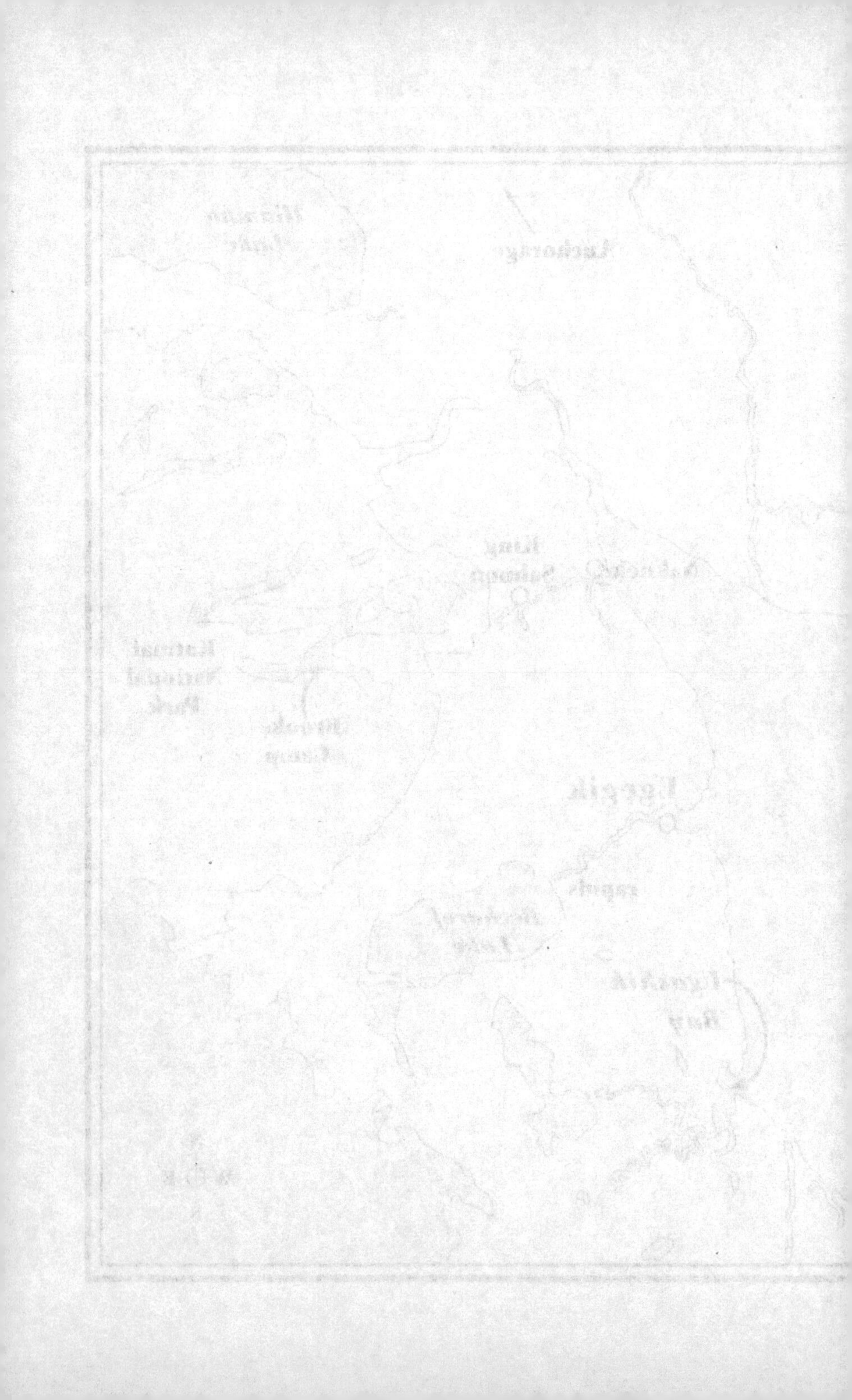

ARRIVAL

I LOOK OUT THE PLANE'S COPILOT WINDOW AND FROM UP HERE the view is perfect and flat in all directions except to the south, where sixty miles away the ground rises in a cone-shaped volcano: the snowcapped Mount Peulik. The sun leans heavily toward the north pole and the land abruptly ends as it disappears into Bristol Bay, which from the air, on a clear day, looks like a flat plate of tinted glass. Looking east and west the flat tundra landscape spreads outward, disappearing at the bend of the earth.

My destination is the small village of Egegik, 350 miles southwest of Anchorage on the western side of the Alaskan Peninsula, a stretch of land that extends out from the mainland 475 miles, and averages 50 miles wide. Cut off from the interior by a vast mountain range, the peninsula is geographically isolated, even from Alaskans. At its farthest point west the Aleutian chain begins, a 1,200-mile strip of islands aimed at Russia in the shape of a kite's tail. The only way to Egegik is by sea or plane, and by sea one must navigate the violent waters of the Bering Sea, not something done by anyone other than commercial fishermen or cargo ships.

I boarded the plane in King Salmon, Alaska, and strapped myself in the copilot chair. The only other seats on the plane were occupied, one with cargo and the other with a female passenger; a Native woman who was busy chatting with the pilot about something. Thrilled by the landscape below I quickly put in my earplugs, and adjusted my sunglasses to shield my eyes from the Alaskan sun, that cosmic torch that taunts all summertime visitors to the Great North.

Now I scan the earth for any clues of a human footprint. A house. A road. A discarded boat, or heap of trash. But there is nothing. From my angle there isn't even a tree, at least one standing over four feet tall. There are no hills, just a flatness, the kind one imagines astronauts see as they peer down at earth from space, the world smashed flat by the relative distance. But there are the thousand shallow ponds that dot the tundra, a broken mirror shimmering the reflection of the plane's metal, exposing how small we are in comparison to the landscape before us.

These pockets of water fill topographical wounds created ten thousand years ago, as the last of the great glaciers slowly receded, scraping the land as they disappeared, like a giant John Deere bulldozer clearing a road. A road the size of Tennessee. As the millennia passed, mountains disappeared, pulverized to rock and pebbles, leaving behind indentations in the earth's surface, which then became lakes and ponds, making the area resemble a gigantic soccer field full of potholes after a fresh rain.

The lack of trees can shock the first-time visitor. The flora is thick but short and bent over, genetically altered by thousands of years of wind blasting down from the Arctic with nearly hurricane force. That isn't to say there is a lack of vegetation. The ground is teeming with green. Tundra grass, alders, and willow squeeze together and cover every square inch of the land. They grow in low thickets, each species intertwining with the next, growing sideways instead of upward.

With the wind behind us, we fly over a river and a tiny village. It looks deserted, not a person in sight, only a cloud of dust rising up behind a single van driving toward the airstrip. There are a few large water tanks, some heavy equipment, and a row of sea cargo containers, but no people. Several large buildings are covered with rusted tin roofing on the bank of the river. Steam billows from a smokestack. This must be the cannery. Most of the homes look abandoned, the grass growing as high as the windows. As we bank I get a closer look at the Egegik River, which spans more than a mile from one bank to the other. The water is muddy, not clear as I had envisioned.

The plane sets down on a stretch of gravel on the bank of the river, just behind the town. A lone orange wind sock stands at attention; the pilot guesses 30 miles per hour, says that is normal out here. There are no buildings at this airport, no small tin shack with the word EGEGIK on it; there isn't even another plane in sight. Instead a van is waiting at the edge of the gravel, near the grass.

The van pulls up to the belly of the plane and we all pitch in, quickly unloading the luggage, along with the U.S. mail. Some groceries and boxes of frozen goods are transferred as well. The driver of the van is a small Native woman with a round flat face and Asian features. Her age is a mystery and she laughs loudly with the young woman from the plane, who also has a round Native face and Asian features. They are talking as if they have been having a conversation for the last two hours. I don't listen. Instead I hold my backpack close to my side, staring out the window as the van begins to move, trying to pick up any clues that will help shape my perception of this outpost. The driver heads down a single-track dirt road toward town.

"Hey, where you going?" the driver suddenly yells at no one in particular.

I say nothing. The driver looks at me in the rearview mirror. "Yeah you, I'm talking to you. Who you working for?"

"Sharon Hart," I say, blurting out the name of the stranger who called me twenty-four hours ago, asking if I wanted a job as a commercial fisherman.

"Sharon's fishing partner is Carl, my husband," says the passenger. "I'm Jannelle."

"And, she's my daughter," says the driver, pointing at the passenger.

DRIVING, WE PASS a few people walking but no one waves, their heads pitched downward as if staring at their feet. I count more four-wheel ATVs than cars, and on one ATV there are stacked several people. It's hard to say whether they are mothers with their children or older siblings with their younger siblings or maybe just friends all riding together.

Finally the van stops in front of a small shack with a caribou rack above the front door. "Sharon's house," says Jannelle, pointing at the dilapidated structure. "There was an opening today, she won't be home for a while. Carl's still on the *Fiasco*, halibuting."

I nod my head as if I understand.

"Get out. And shut the door," yells the driver. Stunned, I don't move.

My first instinct is to compare this place to a shantytown in the Third World. In some ways it does resemble many I've visited. But there is one clear difference. In poor villages, the world over, the local people are almost always friendly. They may not have food on the table or running water, but they welcome you in. And the poorer the place, the nicer they are. At the other end of the scale are the rich, locked behind tall gates, with twenty-four-hour security guards. But here I am confused. Egegik, by the looks of it, is poor, but the people act rich. I know within the first minutes of being here that this will be a difficult place to understand.

"Now!" the driver shouts, waiting for me to get out of the van.

Standing on the road, I look away from the driver, hoping to discourage her from speaking further to me. The passenger leans out the window, raises her eyebrows, and playfully nods her head in my direction.

"Welcome to Egegik," she says, and the van drives away, spitting sand in my face.

EGEGIK

EGEGIK WAS NOT BUILT TO LAST.

It grows out of the ground, in a pitched battle to rise above the tall tundra grass, which swallows everything that is neglected. There are no concrete streets in Egegik, which would only crack during the freezing winters. Instead, wooden planks act as sidewalks, and down by the docks the roads are massive slabs of lumber, making the entire village resemble a decaying pioneer town of the West.

Most of the homes are one story, built from scrap wood left over from some other house that fell apart. Houses have a way of disappearing into the spongy tundra, or decaying from the ice that pounds away at them for several months a year. Rooms are basic, nothing fancy. Some houses have dashes of red or yellow but mostly the wood is either left alone or painted white with a colored trim. In every yard lie pieces of engines, boats, net webbing, snow machines, and wood. And yards are not defined by fences or patches of grass, but instead by how far the collected items extend, usually until a road or a body of water cuts off the forward progress of the junk.

The truth is, no one comes to Egegik for pleasure. The only tourist anyone can remember visiting was a kayak enthusiast who went upriver. There are no cafés, no public restrooms, no rooms for rent, no travel agents or boats for hire. Until the mid-1980s only one phone existed in all of Egegik, located at the liquor store, which is open a mere two hours a day. The only store with food is also open two hours a day, one hour a day in the winter. And a round-trip flight to Anchorage for a family of four costs $2,400.

Out here isolation is not a choice, it is the only option, and eventually it is this isolation that shapes and defines everyone who comes here.

Egegik would not exist, except for one thing: salmon, specifically sockeye salmon. Every year, in June, for at least the last eight thousand years, sockeye salmon, also called reds, enter Bristol Bay. They do not come in the hundreds or even the thousands. Tens of millions of sockeye salmon come, loosely gathered together in the shape of a giant ball, swirling in a counterclockwise motion, resembling an underwater hurricane. Here they will wait for some biological alarm clock that triggers the final leg of their journey. Then, depending on temperature and their keen sense of smell, salmon begin to peel off the ball of fish and enter into the river systems of Bristol Bay: the Wood River, Egegik River, Ugashik River, Kvichak River, Alagnak River, Igushik River, Nushagak-Mulchatna River, Chignik River, Togiak River, and Naknek River.

Egegik, pronounced Ig-GEE-gek, means "Swift" or "Throat" in Yup'ik Eskimo, who, along with the Sugpiaq/Alutiiq and Athabascans, have stomped along its banks for six thousand years or more, as they migrated to the river in the summer to fish for salmon. By winter they all migrated elsewhere, probably back to settlements better suited to fend off the freezing dark winters.

The first contact with white men came in the 1700s, when Russian fur trappers made their way up the Alaskan Peninsula. By 1843 the Russians established an Orthodox church in Nushagak. Soon after, the Russians officially reported a place called "Igagik" as a fish camp. The Russians ruled over this area until the eventual sale of Alaska to the United States in 1867. As they established their colony, they encountered ethnic groups with various dialects and languages, whose loyalties were not bound by tribes but by villages. To make things simpler, the Russians began calling everyone Aleuts, a name still used to describe the people who live on the peninsula.

Today, the Russians' impact can still be felt in several ways. The names of mountains and lakes have unmistakable Russian origins. Also, in almost every village stands an Orthodox church, though often in a state of disrepair. And although many Natives on the peninsula have since converted to the Baptist faith, the Orthodox building always remains a place of reverence for the locals. Furthermore, in Egegik, the church continues to be one of the village's largest landowners. A vast majority of the riverfront property in Egegik belongs to the Orthodox Church in St. Petersburg, which still maintains an office in Sitka, the former Russian capital of Alaska. Few in Egegik still pay rent to the church, but legally, several people who live in town own their homes but not the land they are built on.

EGEGIK THE SEASONAL fish camp became Egegik the village in 1895, once the Alaska Packers Association built a salmon saltery on the north side of the river. But after the town was founded, it had to be moved twice, once because of an outbreak of the plague and the next time for the Spanish flu of 1917, the same one that killed fifty million souls worldwide. The village finally formed at the base of Church Hill, where many of the flu victims were buried.

The town today has a year-round population that hovers between sixty and one hundred, depending on who has died and which young women have been shipped to Sitka to attend the Native school. The one school in town teaches kindergarten through eighth grade and operates only if there are at least ten students enrolled. Any less than ten and the school closes down for the year, and the kids spend the winter at home.

The overall conditions in Egegik can be deceiving, at least on paper. The median household income is $46,000, higher than the national average, but then again the food in the store costs twice as much as in Anchorage, which is already one and a half times as much as anything in the lower 48 states. Add to that the cost of heating oil, gasoline, and the occasional flight to Anchorage for family or medical needs and that median income is equivalent to living in a Third World country. To make matters more dire, almost every family depends on the fishing season for a large share of their yearly income, which explains why three-quarters of all adults in Egegik are "unemployed," a useless word in Egegik, because all it really means is that the locals are waiting out the winter until summer brings the abundance of salmon.

And outsiders. Between June 20 and July 10, anywhere between one and two thousand fishermen will pass through this village.

As the hours pass, I lean against the shack adorned with the caribou antlers and wait anxiously to meet my new captain. It is June 21, the solstice, the longest day of the year, and even though the time is almost ten o'clock in the evening, the sun hovers in a permanent state of golden hour. Bald eagles cruise the riverbank, and the sun pours itself over the ground vegetation, making it seem almost radioactive, as if growing before my eyes. I shield my eyes, swat at mosquitoes, and every once in a while have an overwhelming sense that I should leave while I still have the opportunity. That maybe this was a bad decision.

Finally, I hear what sounds like gunfire or fireworks. I walk to the front of the cabin and peer up the road. Instead of a madman with a gun or kids throwing firecrackers, a 1978 Chevy Blazer comes barreling down the road. It has no windows, no headlights, no doors, and, from the sound of it, no muffler. The Blazer slowly rolls to a stop, passing the front of the house, like a fatally injured tank. A woman steps out of the driver's seat and immediately the vehicle lurches forward, rolling farther downhill. She turns and puts a rock under one of the rear wheels. Problem solved, she walks toward me, wearing waterproof waders that cover her entire body.

"No brakes," she explains, and then grabs a five-gallon bucket from the vehicle, full of water bottles and wet gloves. "Any problems getting here?"

I tell her no.

She extends her hand and gives a solid handshake. "I'm Sharon."

The first thing I notice is Sharon's smile, broad and wide and full of mischief. She strikes me as a woman who has never stopped being a tomboy. Her eyes dart around the yard.

"Seen my kid?"

"Uh, haven't seen anyone."

"Where are my cigarettes?" she asks, more to herself than me. "Is my brother in there?"

I say nothing, not knowing anything she is talking about. She walks past me, over to a sea cargo container attached to her cabin. Curious, I follow her. Inside the container she strips out of her waders, down to a sleeveless shirt, sweatpants, and socks.

I scan the container, trying to inventory all the gear inside. On the shelves are wrenches, sockets, several boxes of screwdrivers, and other tools. Attached to the table is a large vise for crimping everything from car alternators to door hinges. Crowding the cabinets are webbing for nets, and on the floor are anchors waiting to be used or fixed. Raincoats and shin-high rubber boots hang from hooks on the walls. On the back shelves sit cartons for salmon eggs, bolts ten inches long, and washers as large as silver dollars. Scattered about the floor are engine parts: spark plugs, gaskets, hoses. There's a barely visible clothes dryer next to the door, buried under a pile of shoes, clothes, and more tools.

I ask Sharon about the strange shape of the cargo van: its steel casing bubbles out, as if pounded repeatedly by a sledgehammer from the inside out.

"Original owner had some dynamite and ammunition in here and it caught fire," she says, while looking intently under gloves, clothing, and towels. "Blew the thing up from the inside out. I had to jack it up and pour a cement floor in here, but it holds up."

I pause and then inquire what happened to the original owner of the cargo van.

"It's a long story, and I don't know you yet, so I won't go into it. But the short story is he's serving two hundred years in prison," she says, yawning and still searching for something.

I wait for further explanation.

"Double murder. I was the last one to see them before they got on the river that day. Three went out and only one came back. It wasn't too hard to figure out."

She pulls three cigarettes out of a shirt pocket hanging off the wall.

"Found 'em," she screams with elation.

"Anything I can do?" I ask, hoping to make myself useful.

"Can you find my kid?" she says, walking past me and into the cabin.

From the outside, Sharon's cabin looks like every other house in Egegik: a run-down firetrap with shingles clattering in the wind and a roof that leaks. The house is eight feet across at its widest and about thirty feet long, with a sloping floor that runs downward from the front door, toward the river. There is a kitchen sink, but no hot water. The potbelly stove heats the house, and there's a small oven for cooking.

"You got the world's largest toilet outside, but if you need to take a shit you gotta go to camp," she says, explaining the place to me that first night.

She gives me the general layout of the town. Tells me that Egegik is a company town, and always has been. The current bosses in town are Woodbine and Alaska General Seafoods, better known as AGS. Before the season these companies often front the money for fishermen to buy plane tickets, fuel, and food. During the season, if necessary, they prepay for emergency shipments of parts, gear, and gasoline. Toward the end of the season they provide cash needed to pay off the crew. This fronted money is all based on how many pounds of fish a captain has put on the books in the years past and the outlook for the coming season. At the end of the season all the incurred debts are subtracted from the pound total and what is left is the fisherman's profit.

As Sharon already mentioned, the nearest bathroom is at the AGS camp, a half mile away. They also run a few bunkhouses and a laundry room, both used exclusively by captains working for the company. There is a cafeteria that serves two meals a day—charged directly to your captain's account—but the coffee and tea are free all day long. It isn't much, but they do have the bathrooms and, more important, washers and dryers, a pay phone, and hot showers.

Sharon's son, Taylor, ten years old, walks in just after midnight. She scolds him about coming home at dusk and how the bears are out at this time of night. After a half hour of mother and child bickering the cabin calms down.

Finally, Sharon and I sit down over whiskeys and talk about fishing. I tell her I don't know the first thing about it but am able-bodied. She ignores me and talks about nets, engines, missing gas canisters, and people I've never met. She talks about Carl and says he'll be here any day. She speaks quickly and jumps conversations, chasing topics, one after the other, until somehow they all connect. Her fast, choppy sentences hint at some sort of internal code, as if she is speaking out loud a conversation taking place in her own mind. That, or she's in a hurry to be someplace else.

She also repeats herself often, as if she thinks by telling me the same thing over and over I will begin to tell the story back to her. This way of speaking—assuming the listener knows exactly what the speaker is talking about—runs throughout the village, in Natives and outsiders alike. A thread of conversation runs like a river beneath the village, and if a person stays long enough he will suddenly hear the rhythms and remember the names and begin to engage. Like learning a foreign language. At some point, after months of study, the ear, the mind, the lips and tongue begin to work in unison. But until that time, a person must wait and listen, hoping to catch up.

During the first night a few people stop by, each sticking his or her head in the door, shouting out Sharon's name. She answers them all by name and then promptly tells them the party is not at her house tonight. Come back tomorrow night.

"You'll see," she tells me. "This place can get a bit crowded."

A few times I stand, look out the window, and see a person moving down the road, usually a Native weaving from side to side.

I ask Sharon about the locals, the ones who live here year-round.

"It's a tough place to live. The law is at least a day away. The nearest hospital is a plane ride away, so any major trauma and you are dead. You are on your own out here and that does something to people."

"Do they like all the fishermen coming here?"

"Well, let's just say they have learned to live with it."

I WALK OUT the back door to relieve myself. The color of the sky is purple and pink, making me believe sunrise is a few moments away, when really it's just after one in the morning and the sun still hasn't set. Seagulls coo as they ride the current of the wind, which has picked up. Soon the bears will begin their nightly walk along the beach, searching for salmon too weak to fight the current.

I have been warned. By the stewardess, the pilot, the bartender at the Anchorage airport, my family, and anyone else I've talked to in the last twenty-four hours. This is the most dangerous job in the world. People die out there. That's what they said, and they were right. But that doesn't frighten me, not in the beginning. I am too overcome by another feeling. Out here, it takes only a few hours to realize everything is different. This place is ancient, but there are no human ruins. It is ancient because other than the Native people who have occupied this small parcel of land, the view hasn't changed since the end of the last ice age.

Later, in my bed, the mattress sags under my weight and a black mold stain stares down from the ceiling. The temperature has dropped into the mid-forties, but I feel warm. Warm and more content than I was in the overpriced hotel in Anchorage last night. Or the house where I lived for the past three months, with a toilet and running hot water.

The human race is a strange species. We are prone to mood swings. A change in geography can alter one's outlook on life. Two days ago I was picking tomatoes for seven dollars an hour on a farm not far from the banks of the Sacramento River. Then the phone rang and someone asked if I could come to Alaska to work as crew on a commercial fishing operation. I've learned that these unexpected offers have a way of upping the ante in life, as if the mystery isn't deep enough already. And so, I said yes.

RULES

BEFORE TAKING THE JOB I, LIKE MOST, THOUGHT OF ALASKAN commercial fishing as something done in a small tug-shaped boat with a captain's nest, where a cranky old man drank coffee and studied the charts, rubbing his fingers through his thick salt-and-pepper beard. On deck his tired but steady crew stood in thick slickers, either throwing the net out or bringing it in, full of fish.

Those boats do exist. They are called drift boats, and their method of fishing, "drifting," is by far the most common type of commercial fishing. Drifters are allowed to put out up to 150 fathoms—900 feet—of gill net, which then drifts behind the boat. By law each boat measures 32 feet in length, no longer. This rule helps restrict any one boat from hogging the catch.

Sharon runs a set net operation, the other type of gill netting in Bristol Bay. Unlike drifting, set netting is done in a small skiff, not unlike a Boston Whaler, usually between fifteen and twenty feet in length. The size of the operation depends on the captain and how many sites he owns. Some set netters do it alone. Some set net boats have two people inside one skiff, some three or four. No matter what the number of people in the boat, set netters are only allowed to set out 50 fathoms—300 feet of net—from any two sites. The work involves picking the fish out of the net, either by hand or with a tool called a pick. Traditionally made from wood, modern picks have plastic handles that fit into the grip of a hand. At the top, a single three-inch piece of steel hooks outward like an oversized barb.

The routine is a constant throughout a day of fishing. First the net is pulled over the bow and brought to the middle of the boat, then held in place by large pins so it doesn't slide up and down the length of the boat. Standing on either side of the stretched-out net, the fishermen continue to pull the net, a few feet at a time. Once they have pulled a piece of net into the boat with fish in it, the fishermen stop pulling and begin picking the fish out of the net. As soon as that section of the net is cleared of fish, the crew continues pulling the net, allowing the now cleared section of net to drop back into the water while a new section of the net with fish arrives in the boat. All the while the net is still fastened to the anchor system on land and in the water, thus effectively pulling the boat toward shore, inch by inch. The idea is to fish the net quickly and get it back in the water and then do it all over again. As fishermen like to say, "Clean gear catches fish."

The main difference between drifting and set netting is implied in the terms. Unlike drift boats, whose nets are required by law to drift in the current while catching fish, set net operations are stationary, with one end of the net tied to the shore, the other end tied to an anchor somewhere in the river, usually three hundred feet offshore. A set netter must have both a light touch on the engine when setting the net and the ability to crank down on the engine when facing tough currents. And although the drifters take the lion's share of the fish caught, set netters often have a higher profit ratio because they are using a small outboard motor and don't have to pay for boat storage, engine repairs, large gear costs, and overhead.

Each skiff is unique in its design, depending on its owner and when it was built. Sharon's runs twenty-two feet in length and seven and a half feet wide. There are aluminum rollers on each side of the boat, allowing the net to be dragged easily across the midsection of the boat without snagging. Her boat has three sections, each separated by thin aluminum bulkheads, no more than three feet tall. They are short enough to walk over, but the separate sections allow fish to be held in different areas of the boat, important for managing weight distribution on big days. Drift boats hold more and allow the crews to spend the nights on the boat, but skiffs have their advantages—they are quick and cheap, and they can get to shallow water. However, in severe weather they are vulnerable, and every year skiffs are swamped and lost to the river.

THE AVERAGE TEMPERATURE of the Egegik River in the summer is about 42 degrees. The average time a person lives after falling overboard? A few minutes at best, unless a person's waders fill up with water or he gets caught in a net, in which case he will die in a matter of seconds.

Rule number one: On a boat, always carry a small knife on your person. If the net happens to snag a piece of your clothing—a zipper, your shoe, a button, anything at all—you will be pulled overboard. The knife is to cut you free.

OUR FIRST DAY and the engine won't start. Boats speed upriver as Sharon fiddles with the motor, hoping to jiggle the right wire. I stand in the water up to my knees, waiting for her to figure it out. Behind me is Michelle, the third member on board. Michelle is a slight girl with translucent pale skin, as if she's been inside an ice cooler for the last few days. She has a hunched-over posture and speaks in loud outbursts punctuated by shrieks of laughter. Originally from the state of Washington, Michelle moved to Egegik and married Scott Olsen, a longtime friend of Sharon's. Together they live in a shack up the road from Sharon. They have no electricity, no heat, and not much food, and from what I can tell their diet consists of cigarettes and Coca-Cola or, as people up here refer to it, pop.

I am pleasantly surprised that my waders hold off the cold of the water. A fog has settled on the river, making the far bank disappear, which for some reason makes me nervous. Finally, Sharon uses a screwdriver and jumps the charge. There's a choking cough, then the motor roars to life.

Already late, we motor upriver and make a long left turn, taking us directly across the river. Sharon has no maps, GPS devices, or depth finders. She doesn't need them. She almost seems bored on our trip across the river. My eyes are wide, my mind racing with scenarios of how things might go terribly wrong. She lights a cigarette and wipes the hair from her eyes. She never looks down at the water, only ahead, to the far bank, as if fixated on a landmark, although I know she isn't. She doesn't need to be. Below us is an invisible channel, navigable only to people like Sharon, who have fished their entire adult lives on the Egegik River.

The boat bucks against a series of waves and then comes to a stop along the river's edge, a thick layer of clayish mud. The brown water smells like salt, although the sea is seven miles downriver.

"Pull the net out. Lay it on the mud," says Sharon. She maneuvers the boat to point upriver, then kills the engine. I jump out, immediately sinking in mud up to my shins.

"Keep your feet moving, otherwise you will lose your shoes," yells Sharon from the back of the boat, where she's busy grabbing some rope. "Grab the leads."

I hesitate, not knowing what "the leads" are, but too embarrassed to speak up. The entire day I will be hyperalert, my eyes and ears ready to jump at Sharon's every command. I am not worried about failing so much as looking like a complete fool. My overwhelming concern is that she is going to ask me to tie a proper sailing knot, something I never learned to do.

"Grab the line without corks," she barks, seeing I do not know which part of the net to grab.

The lead line. This rope runs along the bottom of the net and is laced with lead, giving it weight, which will make the net, when extended, stand vertical like a volleyball net in the water.

Michelle pulls on the cork line, the top of the net. As the leads pull the net farther into the water, the cork line floats, held up by floating pieces of foam with holes in them.

Michelle and I lay the net near the waterline.

"Can't put it there," yells Sharon from the top of the cutbank, which is where the grass meets the mud.

"What?" I ask, growing more and more frustrated in the mud.

"Move it up shore. Now," she yells.

As Michelle and I pull the net farther up the mud, a low-flying plane passes overhead with the words STATE TROOPER painted on its side.

I make my way up the mud bank toward Sharon, who stands near a piece of plywood nailed to a wooden stake in the tundra. The plywood has Sharon's last name and her permit numbers spray-painted in large letters.

"Fish and Game rules," she says, tying another knot on a rope in her hands. "No part of the net can touch the water before the opening start time." Still busy with the rope, she continues, "Got to have your name printed in capital letters, clearly visible to troopers from one thousand feet offshore. Got to have your name on a cork every tenth fathom. And your permit number has to be on the first and last cork, and on the side of the boat. Any infraction means money and not being able to fish until it is corrected."

Sharon has two sites, side by side, three hundred feet apart. Like a miner with a claim, she retains the right to these sites as long as she fishes them every year, but if she lets a year or two go by, anyone can come and claim the sites. In truth if the site owner does not attempt to fish, anyone can put their nets in that spot on any given day. But the law and local custom are blurred in places like Egegik. And even though it's lawful for someone to fish Sharon's spot because she isn't there, it would take a lot of guts to do it. Why? First and foremost her brother fishes above her and he wouldn't allow someone to sneak into her site. Further, she knows all the other fishermen upriver and downriver, and they would protect her site until they knew she was giving it up.

Behind the sign are several steel rods buried in the bank, each placed a few feet apart with a single piece of rope weaving around them, in figure-eight patterns. The rods have been driven into the ground by a sledgehammer at an angle, aimed away from the river. Sharon grabs the line of rope tied to the rods and ties it to the cork line, effectively anchoring the net to the bank.

"If these ever break loose or you think they are, let me know," she says, and then walks to the water.

"Hey Sharon, do you think we will get them today?" asks Michelle loudly, with a big grin.

Sharon ignores her and tells me to grab the red buoy lying on the beach. When I do, I realize it's tied to the anchor sitting in the shallow water, flukes up.

"Free the anchor," says Sharon.

I walk through the mud, struggling to pick my feet up one at a time. After I untangle the line from the anchor I throw Sharon the anchor cable and she clips the cable to a buoy. Once the anchor is set in the deep water of the river, this buoy will hover above the anchor, allowing us, and other fishermen, to know where it is at all times.

"Every time the tide changes, it fouls the line. Got to check it or the anchor won't work."

Now one end of the net is tied on a running line to the stakes, which during low tides run forty feet from the bank, to the water's edge. During high tide the muddy beach will disappear and the distance between the edge of the water and the stakes will be less than ten feet. What fishermen are trying to capture is the push of fish that comes with the flood, or incoming tide. Because most nets are set in low tide, prior to the flood, the nets tend to dangle in the air, and sometimes between thirty and fifty feet will be exposed while the rest of the net is fishing in the river. Why? Because that portion of the net, dangling for a few hours of low tide, is there to capture the fish that will run toward the shallowest point, the shore, when the water rises. The biggest catches are usually made in this small area by the shore, referred to as the "inside."

Sharon has the buoy, now clipped to the anchor line, tied to a large tow bar rising out of the back of the boat. Using the boat as leverage she slowly pulls the anchor out into deeper water, careful not to let the net leave the shore, for this would be illegal if done prior to the official opening. Instead she unties the buoy and throws it into the water. When the time comes we will grab this buoy and reattach it to the back of the boat and pull the net out into the water.

"Okay, now we do it all over again," says Sharon as I pull myself over the bow, back into the boat, mud oozing off my waders.

At her other site we repeat our steps and this time leave the buoy attached to the boat, waiting to drag it in the water.

"What time is it?" asks Sharon. No one has a watch.

Each day the Alaska Department of Fish and Game announces the time of the opening. These openings can be anything from four-hour windows to forty-eight-hour marathons. Today's opening: noon to 8:00 P.M., an eight-hour stretch. To have nets in the water a minute early can result in a fine of up to $2,500. To have nets in the water a minute after the closing can be equally costly. This explains the plane that flew overhead and why every three hundred feet boats like ours sit and wait, each with a line attached to a buoy ready to drag the anchor and net into the water.

After a few minutes of silence, several boats above and below us race straight offshore into the channel, pulling their nets into the water.

"Now what?" I ask.

"We fish," she says, and guns the engine, slowly pulling the net off the shore into the water.

RULE NUMBER TWO: Listen to your captain.

Sharon has a short attention span and very little patience, and she rarely explains a procedure prior to doing it. She also has a tendency to give instructions in the form of frustrated commands, as if exhausted by telling you over and over the same thing, even if it's for the first time.

But, I remind myself, she's been fishing for twenty-four years. Me? Less than a day.

IT'S BEEN SAID that prostitution is the oldest profession. If so, fishing is the second oldest. And for fishing, like prostitution, nothing much has changed in terms of how business is conducted. Except for the invention of motorized engines and nylon webbing, fish are caught the same as they were two thousand years ago: by throwing a net in the water and pulling it back into the boat. Peter and Paul cast their nets in the Sea of Galilee and fed thousands. Between thirty and forty million sockeye salmon are caught every year in Bristol Bay. The yearly catch of Egegik alone is close to nine million fish.

The management of the Bristol Bay fishery took their cues from the Atlantic salmon fisheries, which are now commercially extinct, overfished by a large fishing fleet with too few rules. When Bristol Bay became an organized fishery in 1972, they made it limited, meaning the Alaska Department of Fish and Game does not issue any new permits. Ever. Between set netters and drift gill net boats there are fewer than three thousand permits in all of Bristol Bay—approximately 1,858 drift gill net permits and 1,000 set net permits.

Even though many captains own their boats, some are owned by a bank that carries the debt. No one works directly for the corporate fisheries; instead commercial fishermen work as independent contractors who sell their fish to the canneries and the fish wholesalers. This separation of labor and management helps ensure each side is playing by all the rules.

To fish commercially in any Bristol Bay district, one must have a Bristol Bay permit, and the only way to get one is either to have obtained it prior to 1972, or to buy an existing one from someone getting out of the business. Each permit holder must carry their permit while fishing, usually around their neck in a waterproof case. Crew members must purchase a crew license for the year, which runs between $100 and $200, also required by law to be carried on their bodies while fishing.

Fishing in Alaska is an expensive vocation. Boats run anywhere between $50,000 and $250,000. In the late 1980s, at the peek of the wild salmon market, permits were selling for upward of $200,000. But since the early 1990s, with the introduction of farmed salmon, the price of wild salmon has fallen, and so has the price of permits. Recently, permits range anywhere from $25,000 to $100,000.

Some boats date back to the 1960s, with small cabins, rusted smokestacks, and outdated engines. These are the less expensive ones. The newer boats have hydraulics to haul up the nets, large horsepower engines, and room downstairs for sleeping. The most expensive boats are the ones powered by jet engines. These boats take the most fuel to run, but they also tend to be owned by the high-liners—a term that refers to those captains who have the largest catches year after year. The jet engines increase their catch because the captain can get the thirty-two-foot-long boat into water six inches deep, which is where the fish tend to run.

It's true the run has diminished over time, and fishing has contributed to that, but overall, Bristol Bay is still the healthiest run in the world. Limited entry and strict law enforcement have played a large part in regulating the fishery. Or, in other words, the real mystery to fishing in Bristol Bay is not whether the fish will return each year. They do. It is figuring out how to stay both financially solvent and alive.

We are picking fish from the net when I hear what sounds like a low-flying jet.

"Luke," says Sharon, pointing to a drift boat barreling upriver. The captain, high in a crow's nest above the boat, waves and Sharon waves back.

"*Quantum Leap.* He'll sit there for a few minutes and then set for the ebb," she says, picking five fish while I struggle with one.

After a few moments Sharon leans over the net and grabs my fish, now stuck halfway through the web, his body crushed by my efforts to squeeze him out one side or the other.

She takes the fish and slams it against the bin. The fish squirts into the bin.

"They aren't antiques," she says.

Over on the *Quantum Leap* the engines fire up and the vessel is on step in seconds. When a boat is "on step" it is going so fast it draws only a small amount of water. Some of the larger jet boats draw no more than six inches of water at top speed. As the *Quantum Leap* skims along a sandbar, the crew runs toward the captain's nest. Then, just as it reaches full speed, the net releases, sailing out the back. Luke, the captain, drives on until the net is almost fully extended and then comes to a sudden halt, the end of the net snapping taut as the boat sinks back into the water.

"If that net caught any piece of you on the way out, it would tear you in half," says Sharon, grabbing another fish and then another. I work as fast as I can, but catching up to her rate of production is just not possible.

The problem with learning how to fish on a fast, cold river in Alaska is that every single moment is a new event, something to learn.

"You got to see the fish," she says, her hands working lightning fast to snap fish out of the net and onto the floor of the skiff. She sees me stuck on another fish, unable to free it.

"It's all physics. You got to see the fish."

IT IS AROUND three o'clock when we take our first break. We have picked the nets twice, each time pulling ourselves to shore, dragging the net across the boat. Up to our knees in fish, Sharon estimates we have 5,000 pounds on the boat and more fish still in the net.

"How much can the boat hold?" I ask.

"Six or seven thousand pounds is borderline safe, but I've put over nine on here before. Almost sank, but if there is no wind, it's possible."

We keep fishing. My arms begin to feel the work of pulling the net. I take a moment and look at Sharon and then myself. We have blood and guts up to our waists.

"What the fuck are you doing?" yells Sharon, looking behind me. I look over my shoulder. Michelle has her gloves off, smoking a cigarette.

"Smoking."

"I can see that. But we are working here."

"I think I hurt my back. I just needed a little break."

Sharon shakes her head, sweat dripping down her face, and looks over at me. Together we reach for the net.

Rule number three: No matter how cold or tired you are, never complain about physical discomfort. You won't be invited back.

Later that night, up in Sharon's loft, we go over the day's work. I am still confused about the process of set netting, the rules, the role of Fish and Game, and about why Sharon has the right to fish where she does but no one else can. I have so many questions. Does she own this shack? Where's her son? Where's Carl, her fishing partner? And all the boats on the water today, they all waved at her. Does she know everyone out there?

It is well after midnight, but outside her window a rainbow comes straight down from the clouds into the tundra across the river, a steel shaft of color driving straight into the ground.

For the rest of the night we sip on whiskey, chased with cold beers. Over these few hours I feel a sensation of equilibrium slowly seep into my brain, floating gently across my own psyche. Why is it that this only happens when I find myself in a place on earth where it's obvious that nature and man have found a way to coexist? I don't mean "balance," a politically charged word when speaking of nature. "Balance" implies serenity and rock gardens and lectures under the last remaining big tree by people who think nature is a benevolent place.

But Egegik? This is different. This is nature challenging my very existence. Up here nature demands respect and attention at all times. Any slip and it will devour me in a moment. I will settle for coexisting with nature. And by *coexist* I do not mean to imply equals; that is a ridiculous notion. Instead, if we are lucky, we are like parasites riding along the dorsal fin of a great white, where the trick of the small creature is to make sure its entire way of being does not interfere with the monster it is riding. That is man and nature.

Looking down, I notice my hands tightening. My fingers have small cuts and my back is sore. Sharon notices my grimacing face as I close and open my hands.

"It will get worse," she says, rubbing her forearm. "I lost the feeling in my wrists almost five years ago."

There is a knock on the front door and then we hear it close.

"Oh captain, my captain, what time tomorrow?" comes a voice from downstairs. It's Michelle. "We're going fishing, right?"

Sharon gets on her hands and knees, looking down the stairs into the kitchen.

"Come by tomorrow around noon, we'll figure it out."

There was no invitation for Michelle to come upstairs and Michelle didn't ask. That is the way it works in Sharon's house: don't ask to be invited. She does the inviting.

The front door shuts and Sharon waits a minute, making sure Michelle is gone.

"She and I did twelve thousand pounds the other day."

"You and her? Or you, and her watching?"

Sharon shrugs her shoulders and lights another cigarette. "She's all right. Never fished a day in her life until two days ago. Just needs a bit of encouragement."

Slowly, Sharon moves toward her bed, obviously in pain. She has put off her surgery for carpal tunnel syndrome until next year. And that was put off from the year before. Her forearms, although chiseled and strong, have long ago melded into a bundle of fried tendons, leaving her with stiff paws that tend to maul whatever she grabs.

"And, she did hurt her back on the twelve-thousand-pound day, which is why I called you to replace her," she says, exhaling her smoke and looking me straight in the eye.

Rule number four: Don't bad-mouth the locals. Sharon hasn't survived this long by making enemies.

SHARON HART

SHARON WASN'T BORN INTO A FISHING FAMILY. INSTEAD, THE road to Alaska for the Hart family began the old-fashioned way: divorce and gambling.

LAST NIGHT WE fished fourteen hours and the last thing I can remember is lying down around 3:00 A.M. I look over to the clock and see it is eight o'clock in the morning and we have an opening in six hours. I turn on the lamp and slowly sit up. All I want to do is walk down to camp for the bathroom and a hot shower.

Then from my bed I notice a man I have never seen before standing in the hallway. He is slender, six feet tall, with a thin mustache. I guess he is in his mid-sixties. He wears a down vest, with a rip on the fabric exposing a flannel shirt beneath.

"Hey, what are you doing? Turn off that light," he yells at me. "Last year it cost a fortune to light this place."

I stand up and click off the light.

"Why aren't you mending nets, or something? She's gonna need some gas. You could cut wood for the stove."

I stand up and put on my pants. I look at the man and sigh. I am in no mood for some strange man to be yelling at me. He's fidgeting with his hands, his face; everything about him is a pile of nerves. I have yet to meet Carl, Sharon's fishing partner, and can only pray this is not him.

"Electricity is expensive up here," he says, almost breaking into a shout.

I put on my shoes.

"Where's Sharon?"

I tell him I don't know but that she may be at the restrooms.

"Tell her I got some pork chops and chicken in the freezer. I'll bring it over. I got cookies, bread." He rattles off the food list faster than I can listen. "Extra mayo, cans of chili, crackers, hot dogs, and lots of meat. Hell, I've even got some extra beer. I don't really drink much these days . . . does the phone work yet?"

"I don't think so," I answer.

He shakes his head in disgust. "I don't know how she does it. I really don't."

Without introducing himself or asking my name, he turns and walks to the door. On his way out he mumbles some things under his breath, eyeballing the mess of the kitchen.

"I wouldn't stand for it, no I wouldn't," I hear him say. "I don't know how she puts up with it."

He opens the door and shakes his head. "Keep the lights off. And this yard is a mess!"

He slams the door shut, leaving me standing in the hallway, confused and slightly angry someone woke me up.

I will soon come to learn this is Sharon's father, Mr. Warren Hart.

So often I want to ask Sharon about her father but don't. Sharon is not a person who readily talks about herself or the interests of her own mind and heart. For that, I have to wait patiently. While fishing, we are consumed by the tasks at hand. And, while partying after a day's work, I am told of the history of Egegik and the famous exploits of the past. There is seldom a quiet moment of sharing from captain to crew. When there is, I listen without interruption.

Over the years, I will collect these moments few and far between and stitch them together to make sense of this person who fascinates me. Usually these moments come while doing mundane work, like sitting on upturned five-gallon buckets in the yard stripping nets. Then, under the endless sky, there is time for her to tell me of her life in Egegik.

And later, as I think back to those times in her yard, they will seem like one extended conversation, four summers long.

SHARON SITS ACROSS from me with a broken net trying to teach me how to repair it. We don't have an opening until tomorrow morning at 5:00 A.M. The net I am working is one we sacrificed on a day there were so many fish the boat had been on the verge of sinking. We'd had to cut the net with our knives just to get it on board. Sharon had insisted on that, bringing the net on board.

"Otherwise that net just keeps on fishing, killing everything it touches," she had said, referring to the danger of allowing a net, once cut, to sink to the bottom of the river. This is not an action many fishermen would even consider. Most would cut the net with a machete and head to the next fishing spot without any second thoughts. But there are many reasons why Sharon Hart is different from most fishermen. Being a woman is just the obvious one.

"When I first started everyone was against me fishing," says Sharon, as she repairs a hole in the net large enough to drive a car through, likely created by a seal bashing into it. "Even my dad was against it. Women on a boat are considered bad luck. Besides they didn't think a woman could handle it."

One day, without warning, when Sharon was nine, her mother came home and put her, along with her younger brother, David, on a plane for Hawaii, where they lived for four years, effectively putting an end to a rocky marriage. As Sharon and David went to school in Hawaii, Warren, their father, sold the house in Vegas, where he had been a casino card dealer. He moved to Anchorage in search of his lifelong dream to live off the land. Instead he became a bartender in a high-stakes backroom poker game.

It was 1970. That summer he met Charles "Junebug" Myers, an Egegik Native who was in town to spend large money from a successful fishing season. After getting to know Warren, Junebug invited him out to crew the following year. Warren has been fishing in Egegik ever since.

"My dad has been fishing for almost forty years. He used to drink. And he was mean. But those days are over. Now he's just trying to make a living and be a good father to us. And he is, always has been, really. But we've had some rough patches over the years," she says, lost in memory.

SHARON HART's first visit to Egegik was in 1979. She was eighteen and worked as a cook for her dad. She made enough money to come back the next year, this time working as a member of the crew. Eventually she became a member in Warren's profit-sharing enterprise. At the age of twenty-one she made roughly $35,000, enough to buy her own boat and build her cabin. Her biggest year was $70,000, in cash. Although Native women have always fished the rough Alaskan waters, Sharon was one of the first white women to operate her own boat in Egegik.

"When I was younger," she says, spreading the net wide to see more gaping holes, "it was like being the beauty queen for a month. I usually had a guy on a tender. A fisherman or two. Maybe one from the beach gang or the night watchman."

Her smile grows wide, her laughter bursting through the howling wind. "Sex was not a problem."

Today Sharon lives most of the year with her husband in the north Olympic Peninsula of Washington State, where she seine-fishes, works as a glass artist, and does occasional tile jobs. She has two children, each from different men. Her current husband, whom she met on the Internet, hates Egegik, and has no interest in fishing or roughing it in her cabin. This is why, she believes, the marriage just might work.

"He has his interests and I have mine. Fine with me. Egegik is part of my life now. Some people take fancy vacations to condos on the beach. I come here. It's my summerhouse. I come here to see friends, to live a different life. And sure, to make some money, but it's more about ownership. Ownership of my own life."

One day we are on the water, about to set our nets. It's a quiet day; the fish aren't running hard yet. In these long stretches of slow work I feel my body acting like a machine, pushing and pulling on the net and fish in perfect rhythm, all the while my mind trying to stay one step ahead of Sharon—what chore needs to be done, or what I'm doing wrong. Then for no reason Sharon suddenly says, "My daughter ran away a few months before I flew up here."

I ask if she wants to go in for the day and get on the phone. She says no, and "besides my fucking phone isn't connected."

We don't speak of it again until much later, as we repair nets under a gray sky and steady breeze.

I act busy, not knowing what to say.

"She came back, after a month," says Sharon, breaking the silence.

"Where was she?"

"All I know is after freaking out for a month I get a call from the police in Philadelphia. After some paperwork and bullshit they put her on a plane back to Washington."

Days later, I ask, "Talk to your daughter?" Seagulls wail in contentment as they sit on the sandbars in the river. The gentle wind keeps the mosquitoes at bay.

"Yeah. Everything is fine. It had been a few days since I last talked to her and I get nervous, like she might take off again."

"So why do you think she ran away?" I ask, slicing through the net, with an overly exaggerated slashing technique she's taught me. Any fisherman will tell you there is a perverse pleasure in the purposeful act of stripping a net. It must be one of the laws of nature: to destroy the thing that inflicts so much pain ends up giving a person a great deal of joy.

"She's seventeen. She just wanted to live her life. I mean I wanted to kick her skinny little ass for running away, but really . . . I was proud of her. You know? I didn't have the guts to run when I wanted to. She did."

I nod my head in solidarity, knowing I can never understand until I have a child myself.

"She showed some independence and survived. I was proud she had those skills. And you know what? Not long after coming home she moved out, got a job, and she's never asked me for a dime. Now we have a great relationship, but that doesn't mean I don't still worry about her."

Sharon's a sturdy woman who walks on land as if she were on a boat: legs planted, eyes steady, and body tilted slightly forward on the balls of her feet, coiled and ready to jump. Solid in personality and body, she reminds me of a tree: you can climb her, even swing from her branches; she will sway but never break. She never slumps or slinks or sneaks into a room. She enters with intention and usually with a wide smile. Strong as an ox, with Popeye arms and hands like meat hooks, she tends to wear tank tops, showing off her generous plunging cleavage. Her strong jaw gives way to large brown eyes always focused on a task, never passive. She is not one to bother with cosmetics or bath products, and her blondish brown hair always seems to be falling down from when she last put it up. Her smile is both genuine and at the same time mischievous, as if always on guard, accentuated by her slightly crooked front buckteeth, which always seem on the verge of falling out. And even though she is well equipped with strong feminine qualities and a flirtatious manner, I believe the reason men have always been attracted to Sharon has more to do with a certain quality she possesses: that she will not and cannot be tamed.

• • •

I FIRST MET Sharon on the phone. She was calling from Egegik and I was trying to cool down from a long day of picking tomatoes. I had told myself the job was temporary, that something better would come through. I was in what some like to refer to as "a transitional phase."

Ten minutes before Sharon's call, a friend of mine, Cindy Rhodes, rang and explained that last year she had been working on a fishing crew in Alaska with a woman named Sharon Hart. She didn't have time to explain anything, only that she couldn't go back this summer due to other responsibilities and that she had recommended me for the job. She said Sharon would be calling, soon.

"If you get on a plane tomorrow morning at six A.M. you can be here in time for the next day opening," said Sharon by way of introducing herself. Her voice sounded exciting and crackled with a cadence of anticipation, as if all things were possible where she was.

"Buy a few things in Wal-Mart or thrift stores," continued Sharon. "Sweatpants, sweatshirts, socks, a warm hat, and gloves, but not thick gloves, mostly just to keep the infections away. You will need cheap tennis shoes, at least one size too big, to fit over your waders. No hooks on the laces, nothing sticking out. Nothing you wear can have anything sticking out. No buttons, drawstrings, or anything like that."

"Why's that?" I asked, my first question of this one-sided conversation.

"The net catches everything."

"Then what?"

"You drown," she said, laughing, but quickly adding, "but that's not going to happen."

"What's the pay?"

"Usually we pay rookies seven to ten percent and make them pay their own airfare. But since I'm giving you such short notice I will pay for half your airfare, but if it's a good year I'm sure there will be a bonus of some kind."

"What's an average?"

"Just depends on the run and on the price they set."

I had no idea what she was talking about.

"When do I have to decide?" I asked.

"You have a few minutes," she said; the phone went silent.

"Okay. I'm in."

This was my first lesson in working and becoming friends with Sharon Hart: it is very difficult to say no to her. Besides having a way of luring you with her passion and determination, she is one of the hardest-working people in an industry dominated by men. She's competitive, not with others so much as with herself. She is known as a person who knows the river and does the work. And in the long run, her legend only grows, and people end up trying to win her admiration, which like a true captain she gives out sparingly.

One night we are sitting in her loft, drinking and telling stories. One of her oldest friends, Mike Deigh, is on the couch. Half Scandinavian, a quarter mutt, and a quarter Native, Mike has blond hair and Native facial features.

"There's no sympathy for a legend," he says, referring to Sharon, who bows her head in embarrassment. Mike's forearms rip with bulging veins, and his right forefinger is wrapped in a thick bandage, smashed by catching it between two boats. On his shoulder is a tattoo of his wife, apparently a voluptuous woman named Vanessa.

Sharon has been explaining how her hands hurt from fishing.

"No way you're getting any breaks here," Mike continues. "Maybe some other woman or my fucking brother. But I'm not listening to your whining 'cause I know it's bullshit. And I'm sure as shit glad I don't work for you."

This is typical of how most people speak of Sharon Hart in Egegik. "Tough" is the first word out of their mouths, followed by a long nod of the head, as if stupefied by the thought. Sharon. Tough.

Another time, a fisherman, lost in a whiskey haze, tells me the same story eight times in an hour. We are downstairs while Sharon is in the loft entertaining several close friends. He leans in and shakes his head. "Whenever I think I've had a tough day I always look out and see Sharon crossing the river with whitecaps. I'm in full rain gear and she's in a tank top. And all day I have this goddamn greenhorn working for me who is complaining about the cold and how his hands hurt. When I see Sharon crossing I turn to him and say, 'You are a fucking pussy.' Hands down she is the toughest man in Bristol Bay."

Fishing has given Sharon an identity, but she is first and foremost a mother, although I don't think she would ever be mistaken for an overly doting one. Instead she is like a mother grizzly with her kids; a provider, sometimes playful, but not always attentive. She adheres to nature's philosophy of tough love; let the children make mistakes and they will grow stronger.

Still, being a mother explains why on any given night during the season, depending on the morning opener, Sharon's summer shack is the social club of misfits and freeloaders alike. Sharon's generous personality, mixed with her ability to play referee, fits the role of den mother quite well. She takes in this crowd with open arms, refusing no one, and tries her best to regulate the liquor while providing food and entertainment for all. That isn't to say she enjoys all the company. She clearly does not, but she doesn't see it as her role to prevent the gathering of people who have nowhere else to mingle. She figures the crowd will work itself out. But there are rules, the biggest being that no one goes up the stairs into her sleeping loft until invited. The stairs are wide enough for two feet, but no more. At the top of the stairs, to the left, is a cubbyhole curtained with a black airline blanket. This is where her son Taylor sleeps, directly under Sharon's high-framed bed.

As for the crowds, they tend to gather in the kitchen around ten o'clock and stay until there is no more rum, vodka, whiskey, or beer. Although no hard drugs are allowed in the house, marijuana, if in town, is passed around like a holy grail.

• • •

As an evening progresses Sharon slowly brings people up the stairs into her room, where the party can easily last until dawn. Being invited upstairs is a bit like being tapped by a college secret society.

In her loft is a couch that seats three comfortably. Her bed is a captain's bed with views of the river through huge glass windows. A bench seat lines the side of the bed, where four more bodies can squeeze in. These are friends, most for many years, two decades or more. Rarely do any of these people have the occasion to see each other during the rest of the year, so to them Egegik is a sort of grown-up summer camp. The party is really a long bull session of men and women exchanging stories and making promises for the following year, which no one intends to keep. Some are poor, with debts too heavy ever to pay by fishing alone. None are wealthy. At best they are scratching out a living. In the loft they barter and trade goods. It is a social market of sorts. Without fail they compare notes on how to navigate in the fog, fish in the high tide, low tide, and rain, which wind is best for fish or worse for fish. And when it comes to sharing secrets of the trade they all lie, knowing full well the others will lie right back. When the whiskey bottle is nearly empty, conversation invariably turns to how the fishery is not managed well, and how the fat-cat executives sitting in Seattle are screwing fishermen out of money.

Many of these nights, I slip downstairs and put in my earplugs. The next morning I always tell Sharon I am unable to consume enough whiskey to follow along, when in truth I sometimes find myself bored by the repetitive nature of the conversation. She smiles and agrees and then has them all over the next night. Sharon never judges them, but it's harder for me. I find myself wondering how they live such extreme lives with no desire to change.

Back outside, fixing nets, but this time it's raining. I can't remember what season it is. This time I talk to Sharon about Egegik, a place that has already begun to fascinate me. There is a dark side here, and I ask for her take on it.

"This industry attracts derelicts," she says, putting the final section of the net on the racks.

"You can enter this vocation with little or no formal education. Don't have to take any exams or pass a test, and if you work hard, have some good instincts, and can bullshit with people, you can make a good living for yourself."

I'm quite sure while saying this she notices my face, which looks a bit nervous. I have been implicated. Found out as a derelict.

"And," she slowly continues, "there are doctors, lawyers, successful businessmen, and even writers. The point is, if you fish, something is wrong with you. There are a million easier ways to make a living."

Every summer, when Sharon leaves she boards up the windows and locks the doors. This is done to keep the animals out. As for the humans, she has another tactic.

"Before I leave I take all the leftover food and any booze, and give it to people in town," she says, moving her arms slowly in large, wide circles, as if explaining a complex theory. "I'm trying to help my friends, but also I'm advertising that there's nothing left in my house."

Three years in a row her house was broken into and things were stolen. Once it was her TV and blankets. One year it was all the food and kitchen utensils.

"I don't blame them. Winters are tough. What pisses me off is when they are done they leave the window or door open and the place fills with snow and animals."

And then there is the time the vandals broke in, stole nothing, and closed the door on the way out. "They broke in and just used my bed. There were stains on the sheets. Someone had used it as a love shack," she says, smiling, a gentle nod to the motivation of the break-in.

I ask if she won ten million dollars, would she still come back to Egegik and live in an ailing shack, put up with the weather, hardship, and the company she keeps?

"Yeah, and that is *why* I'd come back."

She walks toward the house, swatting at mosquitoes. "It's not going to win any home and garden contests, but it's the only summer home I've ever owned."

This is a wild place and it takes a survivor to thrive in it.

THE FLATS

THE FOG FALLS LIKE ASH OVER THE TUNDRA AND DOWN THE muddy banks to the top of the water. Boats are anchored in the current, waiting for the opening.

"Check the skiffs?" Carl, Sharon's fishing partner, asks me.

It's 3:30 in the morning and he's standing over the stove, making coffee. I am next to him, shivering from the cold, putting on a pot of water for tea.

"Not yet," I say, quietly.

"It's the peak. They are jumping. If we're high and dry we're not going to like it," Carl singsongs to me, his understated way of suggesting something.

Carl never yells or orders anything. He suggests. And whatever Carl suggests, that's the best way to proceed. He has a keen sense of both how to stay alive and how to fish.

I liked Carl immediately upon meeting him. Most people do. Carl Adams, known locally as Smiley, stands six feet two inches with a big smile on his face and forearms the size of my calves. Originally from southern California, Carl drifted up to Alaska at the suggestion of Dave Hart, Sharon's brother. It was 1986, and like Dave, he went to work for Sharon's father, Warren.

"Never one free moment," Carl tells me one day, remembering what it was like to work for Warren. "He ran his crews like a chain gang. If we were sitting down he had us get up and brush the nets with boat brushes, to get them crystal clean. Or scrub down the boat every day with solvent. He hated anyone not working all the time. He was always yelling and screaming."

The money was still big back then and payouts were always in cash. Carl tells me the story of how a helicopter would land at the edge of the river, and somebody from Seattle, from the fish company, would step out holding a briefcase. Carl would give him a ride to the tender, where the man would pay fishermen with the hundreds of thousands of dollars in his briefcase. That's how business was done for a long time in the fisheries in Alaska.

In the beginning Carl returned to California after the season, and every summer he couldn't wait to get back to Alaska. At first he ran with a group of young men nicknamed "the lost boys," all of them living in a small shack not far from Sharon. Their nickname came from their over-the-top partying, something that Carl often tells me he's not proud of. When drunk, or on our way to getting there, Carl often finds a way to interject the topic. He always shakes his head, as if thinking of a particular event that happened a long time ago. Something he regrets. "Yeah, those were not my best years. Not proud of it at all. No way."

Today he is Egegik royalty, married to Jannelle, and has two children. His wife works for PenAir (Peninsula Airlines), one of two airlines that services Egegik, and they own a house fifty miles to the east in King Salmon, the airport hub of the Alaskan Peninsula. During the fishing season he lives with his in-laws in Egegik, while Jannelle commutes via plane from King Salmon to Egegik with the kids. All this is possible because Jannelle's parents essentially run the town: they own the only store; her father is the mayor; and her mother is Scovi, the same woman who picked me up on my first day from the plane, and who is in charge of every flight in and out of Egegik.

Carl wants me to check the skiffs. Alert for bears, I step out the back door, through the chest-high grass to the edge of the cliff, thirty feet behind Sharon's cabin. In one hour the beach will be abuzz in activity, set net fishermen getting ready for the day's work. Our two boats are seventy feet below me, just off the beach, caught in a deep swirling eddy that has kept them close to land, but still floating. Perfect. I look upriver.

The gray-blue haze turning cobalt announces the morning. The wind has died and there is no rain. A mosquito bites my hand. Then another. It's small, but worth noting when working in these conditions. The mosquitoes will be a problem, but at least we won't be cold today.

HALFWAY THROUGH the morning's work we have a full boat of fish, just over 7,000 pounds. The pace has been frantic, but it isn't an unusual day so far, just busy. We pull anchor and aim downriver, toward the tender, the large ship anchored in the deep channel of the river. The tender, a large Alaskan crab boat, takes our fish, lifted off by a crane, and adds this morning's delivery to our ongoing tab for the season. Dozens of other drifters and set netters will do the same thing, until the tender's belly is full, usually 180,000 to 240,000 pounds, depending on the size of the boat. By the time the tide begins to ebb, the tender will head out to the bay to deliver the ice-chilled fish to the canneries. On the same flood tide another tender will show up, empty and ready to receive another couple hundred thousand pounds.

Going slowly, we time our crossing at slack tide, easily clearing the bars in the middle of the river. I look over the bow in hopes of seeing them down there, through the muddy water. I know they are there: a ball of fish, the size of a small city. The engine keeps skipping. I think we need gas, and after a few more times of stalling and skipping, I yell back to Carl, "Do we have enough fuel?"

He smiles and takes the cigarette out of his mouth.

"We're in 'em," he says.

"What?" yelling back, not sure what he means.

"In 'em."

Ignoring him I continue with my line of questioning. "But are we missing a stroke on the engine? What's up?"

"In 'em," he says, his face beaming with anticipation. He points into the river.

"Fish."

Then I understand. We aren't out of gas. The engine isn't skipping a stroke and the oil isn't low. The blade of the outboard is slicing through salmon. The mile-wide river is choked solid with fish; every few seconds we hit another. Then another. We pass a boat, anchored in the channel and listing severely to its port side. As we pass the stern of the boat the reason for the listing becomes obvious: there are so many fish piled up in it that the boat is on the verge of sinking. I find out later that the sinking boat's captain put out a call on his radio that he would give away fish to anyone who would come get them.

"Let's take the other skiff to the flats," says Carl, already pumped from the morning's work. Carl is one of those people who actually becomes stronger the more he works. Like a good thoroughbred horse, he needs to sweat and push himself to exhaustion before he even begins to feel any pain. Once thoroughly warmed up he is ready to give his all.

I see only pain in Sharon's face.

She has hardly been able to make a fist all day. But the more Carl suggests we do something dangerous, the more likely Sharon will come back and suggest an even more outrageous feat. This is how fishermen communicate: by bluffing, counterbluffing, then waiting for someone to give in. As a general rule the deeper the pain, the higher the stakes and the bigger the bluff.

Finally, Sharon shakes her head. "Is this what you want to do? Go to the flats? On a day like this? The wind is going to come in."

"Why not?" Carl responds, holding his ground, even though deep down he may want to go home and lie in a warm bed.

I look away, downriver, toward the dark clouds hovering at the edge of the sea. Coming upriver a gaggle of geese flies overhead, honking on their way south, toward land. This indicates a wind pushing inward toward the peninsula. In every direction the landscape offers no comfort. Instead its flatness invites a limitless array of possibilities: We go to the flats and drown. We go and catch nothing and just become very cold. A storm comes in and traps us in big rollers and rough seas.

Checking my body, I know my strength is waning and my sweat is beginning to chill my skin. I have no idea what lies ahead, down there in the dark horizon at the flats, but I have a strong instinct that the smart play is to follow those geese, inward toward land.

Not far away a seal pops his head up, looking for the red buoy of a fishing boat. This is the seal's invitation to an easy lunch. He just swims up and eats trapped salmon out of the net.

It reminds me of the old adage touted by fishermen every day around the world as they explain the essence of fishing: "It's not called catching, and for a good reason. You have to find the damn things."

"Okay," Sharon says and stamps out her cigarette, dropping the butt in the same plastic bag that holds her pack of cigarettes, a lighter, and a few candy bars. It's agreed. She will deliver the 7,000 pounds and then meet us later, on the flats.

The flats are a shallow mud plain, exposed only during low tide. Timing is critical in this area, and only the most experienced fishermen drop nets in the flats. We will arrive an hour before the push of high tide and wait, stuck on a sandbar. Then, as the tide floods, the boat will begin to float. We will step out of the boat and push it farther and farther toward shore, where the fish will tend to run. The gamble is knowing when to stop pushing toward the shore and reverse the process, in order to get out before the tide recedes, which would leave us and our boat high and dry until the next high tide, twelve hours later. Getting stuck can have dire consequences. Once a large cargo boat full of oil got stuck across the river in an area similar to the flats. The boat had to wait an entire month for a full-moon high tide to finally push her off the bar.

The fish don't always come this way, but, if the conditions are right, and the wind pushes them away from the deeper channel, the fish will often be guided to the flats, the shallowest spot in the river. It is the place they biologically sense they can most likely dig a hole, drop their eggs, and spawn. It doesn't matter that they are not actually going to spawn until they reach the lake. They still have the instinct to go through the motion every time the water gets thin.

"Don't put out too much net," Sharon says as we pull away from the tender. "We've only got three hours before we have to get out of there."

Nervous, but trying to hide it, I look at Carl, waiting for a reaction to her warning, but he is long gone, already lost in the bliss of fishing. He sees a jackpot and wants a piece of it, but that's not what keeps a person like him fishing. No, it is the guessing, the tactician in him. He wants to prove to himself that he knows where the fish are today.

Carl opens up the throttle and we head downriver, straight toward the Bering Sea.

After several minutes of heading southwest, toward the mouth of the river, Carl sharply turns the boat 90 degrees and we approach the shore, maintaining full speed. Carl drives the boat as far as he can and then cuts the engine. We drift inland until we stick on a bar of mud. Luckily the wind is still down. Out here the river is four miles wide and a strong wind can create four- to eight-foot rollers, or whitecap waves. With a full load on board this can easily sink a small skiff like ours. A silence now engulfs us. Suddenly all the morning's activity is erased and we sit perfectly still.

Up on shore a coastal brown grizzly bear stands up and takes notice of us. Genetically, these are the same species as the grizzlies that roam Montana and Idaho, but much larger, due to their high-calorie diet of salmon. Unimpressed by the sight of two smaller animals in orange raincoats and oversized tennis shoes, the bear drops to all fours and disappears into the alder thicket. Farther down, on the beach, I spot a bald eagle, no doubt waiting for all the fish that will eventually hit the nets and drop out, leaving them too injured and too weak to fight the tides. And once we, the intruders, disappear for the night, the eagles, foxes, beavers, marmots, squirrels, wolverines, minks, weasels, lynx, and wolves will come out to partake of the spoils. And then there are the seals, already busy eating in the river. The beluga whales will take their share as well. River otters, sea lions, and orcas will hunt under the cover of the muddy water. And let's not forget the birds. The arctic tern has migrated all the way from Antarctica for this feast. The golden eagles, emperor geese, kingfishers, gulls, sandhill cranes, and magpies will all show up as well. Tonight, with unstated thanks, every animal within striking distance of this river will gorge. Everyone except the salmon; they stop eating once they reenter fresh water from the ocean. Now, during this intense drive to spawn, they fast, and for nourishment turn inward and begin to live off their own bodies. First, their scales disappear, sinking into their skins. Then their skins turn from glistening silver chrome to a smooth, slimy, blood-red surface. Finally, as they begin to spawn, chunks of their body drop away. Their collective death is imminent the moment they enter this river, either by fishermen, animals, or their own genetic coding.

The silence is interrupted by a loud noise like a jet engine from an airplane. I turn around and to the south about a quarter mile away, driving parallel to us, straight toward the shore doing 20 miles per hour, is a large drift boat with two massive jet engines spitting a rooster tail of water straight up into the gray sky.

"Captain Jack," says Carl.

The boat slams into the shallow mud and then keeps on driving, sliding across it like some alien spacecraft hitting the earth's crust at a speed-of-light velocity. Finally, a combination of gravity, friction, and mass come together and stop the boat. Then, as if dead, the vessel tips to one side, leaving it effectively high and dry.

"He has jets so his clearance at full speed is less than ours. Maybe twenty inches," explains Carl, anticipating my question. He is getting used to my daily inquiries into how things work.

We are already drifting before we realize the tide has begun to rise and has freed us from our muddy dock. Following Carl's lead, I plop over the side, trying to gain a foothold in the deep mud, which is the color of chocolate and slick like oil. The reason we wear tennis shoes over our waders is because of the mud. One misstep and you can sink up to your knee. The double-knotted shoe is not likely to come off in the suction that is created when the mud swallows all the air around our feet. Most of the time freeing oneself is a matter of shifting the mud by moving the ankle back and forth, but sometimes the grip is too strong.

Whenever my feet get stuck, for even a moment, I am reminded of a terrifying story someone told me on my first visit to Alaska. It is said that years ago a young girl waded out with friends and family during low tide into the mud flats of the Cook Inlet, just south of Anchorage. With a tidal range that reaches thirty-six vertical feet, the tides are second in the world only to the colossal tides of the Bay of Fundy in Nova Scotia. Soon her feet became stuck. At first I'm sure her friends and family joked about her being stuck. No doubt a few of them had also been stuck for a few seconds before freeing themselves again. But something went wrong that day for this young girl. Some speculate it was her small frame coupled with the severe tides that afternoon. In the end all the efforts of her friends and family were no help. By the time the rescuers arrived she was on the verge of drowning from the incoming tide. Knowing from past experience she had only a few minutes left, the rescue team placed a rope around her torso and began to tug her out of the mud. In the end she was physically torn in half, killed by the mud that would not release her and the helicopter crew, no doubt equally determined to save her life. Whether the story is urban legend or fact does not alter my sense of fear and dread every time I step foot in Alaskan tidal flats.

Mindful of the mud beneath us, we move the boat several more feet toward shore, until we hit a high spot. We wait a few minutes before another high flow frees us again. With each surge of tide we move closer toward shore.

Finally Carl decides we are close enough. He grabs the sledgehammer and the sign he brought along—his name written in spray paint across a three-by-three foot piece of wood nailed to a driving stake. He walks the sign onto shore, keeping one eye toward the tundra where the bear disappeared twenty minutes ago. Once Carl reaches the grass he pounds the stake into the ground. We are now legal to fish this water.

As Carl makes his way back to the boat I sit in silence, wet on the outside but warm inside my suit. In the quiet I take stock of my aching body. My arms and torso throb from the morning's load.

Still, the bloody cuts in my finger joints and the strained tendons cannot subtract from my state of being, which seems to thrive in the harshest conditions. No longer am I thinking in terms of hours and minutes or mornings and afternoons, but instead I am caught up in the rhythms of this place. There is no rent to pay, no phone to answer, no mail to read. There is no television to stare at, no news to muddle the brain. There is nothing but high tide and low tide, the wind, and the sun. The ebbs and flows are the new clock. This is the clock of our ancestors, an ancient instinct given to us at birth but denied us the longer we live our modern lives. I am running high on anticipation of what may come. The greed of Alaska is speeding through my veins. The call of the Klondike is upon me. This is the day the river opens up and spews out gold, red gold.

We wait until the tide is right and then slowly begin setting the net over the side into the shallow water.

I want to laugh out loud, but instead shake my head in wonder. By traveling to a remote fish camp and working until exhaustion I have escaped the thing that plagues modern society: time, or more accurately the impossibility of ignoring the passing of time in one's mind.

"Velvet time," says Carl as he climbs back in the boat.

Carl reaches into his dry bag and pulls out a bottle of Black Velvet whiskey and two candy bars. Hungry and slightly light-headed, I leap at the chance to take a shot of whiskey, but also want to abstain, nervous of losing my edge for what may come.

"Think the fish will hit out here?" I ask, taking a shot, the only real choice I have.

"Never know about the flats."

Carl's ears are evidence of his life on the water. Chapped and burned, the top quarter of both his ears have wilted, the skin peeled back and then callused over. I once asked if he'd ever been to the doctor about it. "They would just tell me not to fish as much," he'd said. But he did say he was trying to take better care of them, by adding sunblock.

Carl is everyone's go-to man. When he walks into a room the party picks up. When there's a problem with the boat, truck, engine, or anything that has a spark plug Carl is there to fix it. And always he is happy about this. One time I asked why he chose to live in such a harsh environment. He answered, quite simply, "I don't need money to live here. I can barter almost everything. And that makes me happy."

A splash. Then another. And another.

"Freshies," says Carl, referring to fresh hits on the net.

Another one. And then a dozen hit in less than ten seconds.

"Here they come," says Carl as he stands in the skiff, balancing himself with his sturdy legs. He stares out at the net, now bobbing with salmon trying to thrash their way free of our web.

It all happens so fast and with such fury I barely remember how it began, but I will never forget the next few hours. Within minutes the noise of salmon slamming into our net becomes so loud Carl and I have to yell at each other to be heard. Immediately I break out in a sweat, filled with fear of being overwhelmed by the fish, and at the same time feel a huge dose of regret at taking that slug of whiskey.

"We shouldn't have put out so much net!" I yell, in a growing desperation, and remembering Sharon's last words.

"Yeah. Start picking, fast!" yells Carl with some urgency.

It takes one tug to realize the net is so heavy with fish we won't be able to overhaul it across the boat. Carl jumps in the water and grabs three feet of net and gets under it, resting the net on the width of his back, and using his legs he pushes the net up and over the boat's ledge. So while I pull from above he heaves from below. In three feet of net there are more fish than I can count. Every single hole in the net is plugged with a fish fighting for its life. This three-foot section weighs up to two hundred pounds. Carl and I manage to smile at one another, but this will soon turn to panic.

"Pull the rollers in," yells Carl. I move to the front of the boat but slip on some salmon guts and fall into a pile of fish. Quickly I get up and unscrew the roller pins, which lowers the edge of the boat and makes it easier to push the net inside. I rub my hand on the small of my back, feeling the ache of the fall. Carl yells for me not to lose the roller screws. "We'll need them to make it back." I don't know what he means, but I frantically put the pins and screws in the dry compartment under the steering wheel. Within minutes there is a red and brown pool of gurry—the blood and slime of salmon—on the floor of the boat.

For the next hour we use this method: one person in the water shoving the net inside, while the other holds it steady from within the boat. Then the person in the water jumps in and together we pick the fish out of the net. My fingers are covered in blood; some of it from the fish, but mostly mine as the net has begun to rip the skin off my fingers.

Every five minutes or so we alternate who is in the water pushing the overflowing net into the boat, one foot at a time. The noise of the splashing fish is deafening. In the water salmon are banging into my legs, knees, and ankles with a ferocious energy. They are flying between our legs and shooting out of the water toward our faces. Carl and I have stopped talking, trying to use all our energy to pick the fish from the net. But picking the fish is taking too much time. Besides, my arms and fingers are now seizing up. The pain would be unbearable if there wasn't so much utter panic. And in that panic I have done something fishermen do not do unless faced with imminent danger to life or property. I have used my knife several times to strategically slice the net, hoping to free a fish. At one point Carl sees me, but he doesn't say anything. Meanwhile I watch him, trying to learn. He picks the fish out of the net using muscles I'm not sure I have. His fingers are the size of pickles and his forearms like cylinders of steel. He shreds through the net while I try to work around it. We have only seconds per fish; otherwise there won't be time to get off this bar.

I recognize this moment from my dream. The gray sky. The nets full with fish, and the frenzy that allows me to go overboard. If I were asleep right now I would be sweating through my blankets.

But my main fear is not drowning. That is about third on my list. My fear is of not being up to the task, of not having the strength or the stamina to do this job. Any person who has worked in a crew knows this feeling. In any job one can feel this anxiety, of being the weak link, letting others down. This fear plagues me when it comes to fishing. I am not scared of the work, only unsure whether I will fail in trying. Further, I'd be lying if I didn't say I am scared on a deeper, more primal level, of the salmon's desperate energy to live. That, combined with the overwhelming number of them, is beginning to make me feel as if we have been outmatched.

The boat is now almost full. Carl tells me we must pull the net out of the water and get out of here.

"Where is Sharon?" he yells toward Egegik, which is a distant beacon on the bank, a couple of miles away. "Closing is in two hours."

"Does she know where we are?" I ask, catching my breath and trying to get the slime and blood off my face. I ask the question with all the sincerity I can muster, but really I am just trying to buy time from pulling the net. My muscles, from my neck to my big toes, are depleted. I am soaking wet under my waders, not from the water, but from sweat, and my head is getting light from lack of sugar.

I sit back on a pile of fish, some moving around beneath me, gasping for their last breaths. How many have sacs of eggs, mothers desperate to get rid of their full bellies? How many salmon have we denied being born? From the first time I pulled in a net of fish I have been struggling with a part of this job that just feels flat out cruel, maybe even murderous.

In this brief interval I suddenly notice something is wrong: my hands have stopped working. I can barely move my fingers, and the muscles in my forearms have seized up. I look at Carl and he knows I am useless. He stops picking fish, reaches into his wet bag, and gives me a candy bar. He tells me to rest for a few minutes. The pain is building and my exhaustion is beginning to cloud my ability to see our way out of this. The temperature has dropped several degrees, allowing any gust of wind to cut through the rubbery wader material and send a chill through to my sweaty skin. Hypothermia will set in soon if we don't get out of our sweat-drenched clothes. To make matters worse the wind, like the salmon, is riding the high tide in from the sea. It begins to blow.

"Sharon will be here soon," says Carl, almost to himself, as the noise of the fish hitting the net around us only grows louder.

Moving around inside the boat is not an option. At one point I lose my shoe when I try to free myself from a pile of fish that has gathered around my feet. There are so many fish in the boat that to go from one end of the boat to the other we jump into the shallow river. And walking through the water becomes increasingly strange, with the fish banging against our legs like blind lunatics trying to flee a burning building.

Here is the real problem. Our boat is now too full to take any more fish. The tide will turn again in a few minutes, dropping, and our net is still in the water, full of fish. If we have a net in the water at the time of closing, Fish and Game, who have already flown a floatplane above our location, can pull our license until we pay the hefty fine. They won't wait until the end of the season to let us know; they will either land a helicopter on the water or run a twin-engine Zodiac out here to tell us.

Moreover, if the boat is too heavy we won't make it off the bar. Then we will end up sleeping out here until someone comes to get us. To top it all off we will likely lose the load we already have because the fish will go bad overnight.

And then, as we lie on the fish, staring upward, hoping for a solution to come to us, Sharon's skiff appears. She steers close to our boat and cuts the engine. She sees the exhaustion and desperation in our faces. She immediately jumps in the water, getting wet up to her waist.

"Aw fuck, Carl, I told you we shouldn't have done this. We could have taken the same amount from our sites."

"I knew you were going to say that," laughs Carl. *But* he is laughing.

"What are we going to do?" asks Sharon as she leans her massive forearms on the edge of the boat. Her hands look like bruised animal paws. It's obvious by the way she is holding her hands that she can't pick. Her tendons are frozen and she can barely make a fist.

"Closing is in an hour and a half," says Carl.

"Start loading into my boat and I will close the net," says Sharon.

Quickly, Carl and I bring her boat under the net. We pick while Sharon stands in water up to her waist, forty feet behind us, deftly sewing the net shut. She has a hanging needle in one hand and the net in the other and she is tying the top and bottom together with the webbing. This will allow us to be legal if Fish and Game find us after closing. The rule is that a net can be in the water as long as it is not open, meaning it is sewn shut every two feet. Theoretically this makes the net inoperable. After several minutes the boat is half full and we are barely one third of the way through the net. The fish are still coming.

"Carl! We have to bring it all in," barks Sharon from the water, trying to shout over the salmon attacking the net all around her.

"You sure?" he asks.

"Look at the tide. It's changing. We have to get out of here."

Carl motions for me to jump out of the boat. We attempt to push it away from shore. It doesn't move. We reposition and heave with our backs against the boat. We heave and heave and finally it budges, but not enough. We count to three and do it again. It frees up and flows a few feet and gets stuck again.

"Start bringing the net in, we'll deal with it later. We got to get out of here," yells Sharon. She drops what she is doing and comes back to help us pull the net into the boat.

My arms are giving up. I have no more strength; even my adrenaline is waning. The ache is overwhelming. "Come on, Bill, you're doing great," yells Sharon, like a coach egging on her player in the final push. "Do this and you pass the test. It doesn't get harder than this."

We are all giving up, but somehow between the three of us we find enough strength to push, and at the same time pull the net into the boat.

Until we start piling the net on top of itself I don't realize the extent of what lies ahead. But after seeing the massive coil of net, full of fish crushing one another, I realize the problem. We will have to find a way to untangle the net and get the fish out. When I ask Sharon how long that will take she grimaces and says, "All night."

But first we have to get out of here.

"We're stuck again," says Carl, quietly, whispering the words.

For half an hour all three of us slush through the mud as we push the boats, by hand, through the shallow water. Every ten feet or so one of the boats hits another small bump in the bar, stopping our forward movement. We aim toward the deepest part of the channel, more than a thousand feet ahead of us. Each time we stop we all get behind the stuck boat and shove through the obstacle.

At last we have a little over two feet of clearance below us. Carl gets in one boat and Sharon and I take the other. Closing was an hour ago, but finally we begin our journey toward camp.

"WELL, THAT WAS HARD," says Sharon, "but this is where it gets dangerous." She has both hands on the wheel, eyes staring straight into the raging river ahead. I turn back toward the flats to see Carl, not far behind, also going very slowly toward camp.

"We get a wave over the bow and we will lose the boat. This load will sink in seconds."

"Ever happen to you?"

"Not here, but I got swamped by our sites once. My brother swamped his boat just around the corner up by camp. Lucky for him it was in shallow water."

Dressed in a tank top that reveals her meaty arms and hefty bosom, Sharon eases back on the throttle to make the bow ride the crest of the waves, keeping it inches above the waterline. At that moment I can't help but wonder how many men have underestimated Sharon's abilities over the years.

"Oh shit, here comes our nightmare."

To our left is a drift boat and it's flying through the channel, in a hurry to unload fish at the tender. That's the thing about big fishing days. You are never alone. When everyone in the fishery has huge loads of fish, then everyone is facing various degrees of crisis, depending on their gear, their crew, and luck. This drift boat probably spent the last twelve hours on the water and its crew wants to unload and get ready for tomorrow. With repairs to nets, filling up with gasoline, eating, and maybe sleeping there isn't much time to prepare for the next round.

"He doesn't slow down he'll dump us," says Sharon.

The boat is at least three hundred feet to our left but still any wake bigger than a half foot can swamp us. There is no way to signal to him or to ask him to slow down. This is open water, and although he has a legal obligation to slow down for us, he could also continue at full speed and, when asked, say he never saw us.

And then, as if he has heard my telepathic pleas, he slows down. More than likely the captain recognizes Sharon. Respect out here goes a lot further than the law or a sense of responsibility. It's like living in a small town. What one does to the tourist passing through is different from how one deals with locals, neighbors, and those who make the fabric of the town. You have to respect them or else it will come back to you. If he had sped by and thrown a wake strong enough to sink us, there would have been ramifications from within the community of fishermen. And that isn't worth the price of saving a few minutes of that captain's day.

Slow and steady Sharon drives us toward the AGS sign on the town dock. We are almost there.

"Guess you just paid for your flight," she says. "And a bit more."

TWENTY-EIGHT THOUSAND POUNDS

BY THE TIME WE LAND THE SKIFFS ON THE BEACH IN FRONT of Sharon's shack it is 11:30 P.M. After a quick change of clothes, a few sandwiches, and a cold beer for each of us we return to the skiffs, both full of fish. The tide has ebbed, leaving the boats dry on the beach. We strap on our headlamps so they rest just above our eyebrows and get to work picking fish. Looking up, all I can see in both directions are dancing beams of light; everyone got slammed tonight and everyone has on headlamps to guide their way through the black of night.

Trucks and ATVs are roaring up and down the beach. Now it's two in the morning and it feels like the hustle and bustle of a major port. Dozens of drift boats are tied off the backs of the tenders just off the beach. The crab lights cast a bright orange glow on the river. The cranes run nonstop, lifting the fish off the boats. The seagulls are still squawking but it's hard to hear them over the sound of engines, pulleys, talking, arguing, and occasionally some laughter.

Farther down the beach we hear some yelling at the Ivy site, but it isn't clear what is happening. It's too dark to make sense of anything. All I see are small dots of light running up and down the beach.

The Ivy family is run by the patriarch Charles, originally from Texas, but now related to a local family through marriage. Charles's gut hangs far over his belt line, which is held up by suspenders decorated with American flags. He's done his share of fishing, but these days he doesn't lift a finger toward the hard labor. Instead he is the boss of a large commercial operation: six sites in all and a crew of fourteen, including the drift boat crew that works a separate permit. For most of the season Charles can be seen slowly driving up and down the beach, using one of his four muscled-up Ford pickup trucks to pull in nets, push skiffs in the water, and haul gear.

Suddenly, out of the dark David Hart, Sharon's younger brother, approaches on his four-wheeler ATV.

"Hear what happened to the Ivys'?" David asks, baiting our reply.

"No, David, but I have a feeling you're going to tell us," says Sharon, busy picking fish and favoring one arm over the other. David likes to watch others deal with what he believes is misery that he avoided due to his expertise.

As for the excitement at the Ivys', it seems Charles didn't engage the parking brake when he stepped out of the truck. The truck rolled into the river, with the engine running. For a while it was visible, then it began to sink. With a minus tide expected—or when the ebb tide is larger than the high tide—the shift in the water tonight will be extreme. The river could bounce that two-ton truck halfway to the Bering Sea before sunrise.

"You gotta pull it out of there," David yells at me, pointing at a fish I am struggling with. He always yells.

"Right there," he barks, pointing to something in the net, but I have no idea what. "It's gonna take all fucking night the way you are doing it."

I continue to ignore him. Sharon takes a breath and stares at David.

"What?" he says, arms open in a defensive posture. "Just trying to help your crew do their job."

"Davey, Davey, Davey," says Carl, chuckling to himself.

David Hart is a person who doesn't hear other people speak because he is too busy dishing out advice. It's not uncommon among the fishermen I've met. They want to prove how much they know or, looked at another way, how much you don't know. I once asked Carl about this and his response was simple. He said, "This is all they have."

Time passes slowly as we pick, while also swatting away mosquitoes. Then Dave notices something sticking out of the water, not far from our skiff. It's the antenna from the lost truck. Carl jumps on the four-wheeler and drives toward town. By now a crowd is gathered, just in time to see Carl arrive with the tractor, driven by Bobby Deigh, the village heavy machine operator. Carl quickly lashes one end of a thick chain to the blade of the tractor. Meanwhile John Ivy gets in a skiff with Dave and ties the other end of the chain around the now exposed roof and in through the driver's-side window. The cab is half full of water, but luckily the truck is floating right side up. After securing the chain, the tractor slowly backs up and soon enough the truck is pulled from the river, dripping wet.

An hour later, Carl returns and we are still picking fish.

"I fucking saved their ass," yells David, now standing over our skiff like a pumped-up gladiator. "Saw the top of the truck and called it. I don't know if they can ever start it, but we got it out."

"That's great, David. You did a good thing," snaps Sharon without looking up. Carl, saying nothing of his own efforts, laughs a bit and keeps picking fish.

"What? I did," says David.

Since leaving this morning we have been on the job for twelve hours. We have several more hours to go and preserving energy is important.

"Yeah, figured I did my good deed for the day," Dave says with great pride, still yelling and still, it seems, waiting for a round of applause. "Fucking just saw the tip of it and we got a chain on it."

In some ways it is easy to understand David's need for approval. For I am new to these people, this work. I am also looking for approval or, more accurately, looking to make sure there is no disapproval of how I am doing. I have a fear of screwing up, of endangering other people, of panicking on the water. David, who has fished here for more than twenty-five years, the first ten years with his father, is a forty-year-old boy looking for some basic nod for a job well done. Well, he isn't going to get it from this group. Fishermen don't give one another compliments. They are careful whom they work with. They rarely complain of the cold weather or the lack of sleep. They eat terrible food and drink too much caffeine and booze. Friendships are based on seniority and experience on the water. And even then fishermen can be ruthless to their friends.

Three years later, on a night of heavy whiskey drinking, Carl speaks of this night on the flats. He is telling a rookie about a big day he had at the flats, one of his biggest single days of catching fish. He says, "Bill hung in there. Not many people could have." I nod like a veteran and down another shot. I don't dare show my true feelings, that I am jumping up and down that Carl has seen me in this light. In that moment I realize I am more like David than I want to admit. Fishermen don't pat one another's backs. They just tell the story.

It is 11:00 a.m., the morning after picking, by the time we deliver the final fish to the tender. We have been at it for almost nineteen hours. In the end we deliver 28,000 pounds of fish, 16,000 of that from the flats.

"So you will be back next year?" says Carl.

"Ask me in a few days," I say, eating my third sandwich in ten minutes.

In 1989 this same catch would have been a $50,000 day. But today the price of salmon hovers around forty cents a pound. My take is 10 percent, or $1,120 for nineteen hours of work.

"Get some sleep. The opening is in a few hours," says Sharon as she climbs the stairs to her loft. She smiles, exuding a maternal instinct, checking on her flock. She's tough and demands a lot from herself and others, but she is fair and most important she has conquered the most necessary skill in the fisherman's world: knowing how to adjust to the situation.

I pop 1,600 milligrams of ibuprofen and two codeines, and chase it with a shot of Carl's Crown Royal. I dab some Super Glue on the cracks in my fingers, created by the nets tearing into my flesh. In the first few days Sharon gave me a stick of glue, said it was the quickest way to stop the cracks from growing too large.

Carl trots upstairs, in his socks and sweats, holding a beer. He will sleep on the couch. There's no need to go home and disturb his sleeping family. They will never know what he did today, not really.

I draw the blanket over the entrance to my room, shielding me from the Alaskan summer sun. Horizontal, I think of the ancient Alutiiq method of killing whales. They used the poison from the monkshood plant (*Aconitum*), a blue-flowered herb that grows in meadows throughout Alaska. The Alutiiq people dried the roots before pounding and mixing them with water. Then they added the preserved remains of a dead whaler to the concoction to give the drug spiritual potency.

They shot the whale with a dart dipped in this toxic mixture, usually in the tail. Death was not a result of the toxin itself, but instead the drug lowered the whale's body temperature and blood pressure, and paralyzed its nervous system, which led to death by drowning. The death of the whale took up to three days, plenty of time for the whaling party to consider their relationship with the whales.

Whoever shot the whale would return to the village and lie down in his home. There, for these long three days, he would wail in pain and play the spiritual part of being the mortally ill whale. People would gather and pray. Meanwhile, out on the ice the rest of the hunting party would scan the horizon, looking for the whale. The whole purpose of the man faking illness in the village was to pay homage to the whale. He was the whale's dying spirit and the prayers and his acting were part of a ritual to guide the actual dying whale to a beach where they could find it. They were calling it home to die.

I think of this and can't help but feel as if I have cheated today, stolen something from the wild without a return sacrifice. I hurt, but is that enough? Where is the salmon god? What price will I have to pay? I have looked into the eye of the salmon today, seen them gasping for their last breath, fluttering in their death.

Then my internal debate shifts. I tell myself I am a fisherman, an ancient stamp of citizenship in this world. That is worth something. People all over the globe will nourish themselves with healthy doses of salmon, rich in omega-3, commonly known as brain food. I feel a twinge of pride that I am a small speck in the process of feeding our species.

Then again, I think of the salmon. We have killed more than five thousand fish in one day. If hunting can be justified by how much is needed to feed a person and his loved ones, then in one day I have killed enough fish for a lifetime.

These thoughts are overwhelming, so I turn my attention to the most immediate emotion: a satisfying feeling of hard work.

Upstairs I hear Sharon, moaning in pain. She will do this all night, and most nights this season. Her arms are permanently numb from carpal tunnel syndrome and she sleeps sitting up with her arms resting on pillows. For medication she takes an assortment of painkillers, none of which rid the pain, but sometimes let her sleep.

I flex my fingers, trying to stave off the stiffness that will come the moment I fall asleep. I fight like a lion against sleep, wanting to avoid the reality of actually waking up. Or dreaming.

WITHIN A MATTER of seconds, or so it seems, Carl is tapping my leg. I have been asleep for two hours but all I remember is the dream: a flurry of fish getting into the boat, the noise, and panic of drowning.

In the kitchen Carl grinds coffee beans. Above me Sharon's feet hit the floor. "I don't want to do this anymore," she yells downstairs. "I quit."

"Okay. Let's quit," Carl says loudly, followed by a laugh. Sharon mumbles something but it doesn't matter. These two can't quit. The crackle of the fire in the potbelly stove helps to wake me. I stare at the ceiling and ready myself to get up. Sitting up I realize my arms from my elbows down are completely numb.

"Check the skiffs?" says Carl as I pop my head out of the room. I put on my hooded sweatshirt and boots and stagger out the back door. The nets have to be in the water in forty-five minutes.

A soft pink sunrise welcomes me into the day. What day? I have no idea. I lost track long ago. The only choice left is to forget about time and instead let the silent clock of the sun and tides bleed the days together, like an endless red summer.

FISHERMEN BLUES

THE MEN STAND AT THE DOCKS AND TALK ABOUT BELUGA whales and in the next breath a place called Ugashik, and how they are losing their fish to both. These are veteran fishing captains, each dressed in a version of the same clothes: sweatshirts, sweats, and slip-on shoes with no laces. Most are over forty, wear baseball caps, and no one has shaved in a few days. They ignore me. Fishermen, as a group, are like the mob: if someone within the community doesn't introduce you, you don't exist to them. You are a civilian. So, as a civilian, I stand and listen.

"The fucking Japs are ripping us off."

"And the Russians."

"We should strike to get the prices back up," one says.

"A strike. Fuck that. Look how good it worked last time. People still fished and pretended like they weren't. No way."

"Point Moller said a swarm of Egegik fish passed by yesterday. Should be here by tomorrow night."

"They'll keep us closed. Set netters don't have enough yet."

Here comes another, this one with a ponytail, mangled teeth, and a shirt undone to his navel. His jeans, covered in grease stains, have a hole where his pecker meets his thigh.

"Son of a bitch just keeps driving it the wrong way. I tell him and he don't listen. Now what? I'm on the dock fixing a blown engine. For the second time in two seasons. Motherfucker, I hate when my boat don't drive just right."

This is Vince. Declared hobo of the fishing trade, Vince doesn't own the boat or the engine he is speaking about. He works as a crewman and almost every captain knows him. They know he can fix an engine in the dark with ten-foot swells. They also know if you hire him you have to suffer his wrath.

I walk along the dock and see more men in sweatpants, slip-on shoes, and ball caps. There are a few women, mostly working as crew on their husbands' boats. Many Native women fish, but they tend to stay away from the docks. Away from the outsiders and their loud tendencies.

The captain over there stands tall and looks rich. And he is. He is famous for treating his crew like dirt. He charges them for all the food they eat; he even takes a piece of their percentage for the gas used to catch fish. One time Sharon invited his crew over for dinner, to give them a good meal. They told us he gave them each a can of peanut butter in the beginning of the season, and since then he hadn't given them anything for free. No crews work for this captain for more than one season.

But, regardless if they like their captain or not, crews usually stay for the season. The prohibitive cost of getting to Egegik has indebted them, turning them into indentured servants.

When not fishing, sleeping, or getting ready to fish, I often wander down to the docks to listen to fishermen tell stories as they keep a watchful eye on their boats, tied up to one another down below. Like a dog, each boat resembles, at least in personality, the person who owns it. The man who steps out onto his deck with holes in his sweats, a rat's nest for hair, and yelling at no one in particular is standing on a boat with cracks in the fiberglass, frayed ropes, and a broken smokestack. The man whose wader boots don't even have a scratch and whose jacket has a zipper that works will have a crew on deck busily wiping the windows of his captain's quarters.

Much of the season the drift boats are tied together at the dock, bouncing into one another as the captain and crew sleep down below. Strapped together, they are all waiting for the voice on the radio to announce the next opening. Many times I've been in the cafeteria reading three-day-old newspapers—which arrive daily from King Salmon on the plane for the cook—when a captain bursts in the room yelling for everyone to untie from his line. He needs to get out. No matter how much another captain wants to continue eating, sleeping, or ignoring the cry to move, they all rise, go down the ladder, get on their boats, and move. Everyone complies. There is no choice. One day he may be the one who needs to get out.

The names of boats are important to fishermen. Once a boat is named it can never be changed. That is, unless you want bad luck.

The names all have a story: *Foxy Lady*, *Naughty Lady*, *Two J's*, *Yakatak Eagle*, *Vicky Lynn*, *Sheryl Lynn*, *Fiasco*, *Ocean Magic*, *Captain Jack*, and *The Predator.*

Some drift boat names are truly inspired, like the *White Cane.* The first time I see the *White Cane* I am standing on a tender. As we deliver our fish, I look out and see a drift boat closing fast, no more than a hundred yards from Sharon, who is in the skiff below me, attaching a bin to the crane that will bring it on board the tender. The *White Cane* looks like it is going to ram her.

I yell down to Sharon and she looks at the boat and then up to me. "It's the *White Cane.* They know what they're doing."

On the flying bridge stands a man staring straight at me with very large binoculars. The boat is now within fifty feet, and closing. It seems impossible someone is using oversized binoculars at this close range. Again, I yell down at Sharon to watch out. She ignores me.

In a few seconds the boat slows down to a purr and sidles up to the tender, never once touching our boat or the tender. A few minutes later Sharon introduces me to the co-owners of the *White Cane*: Randy and Billy, two brothers, one from Idaho, the other from Washington. They have been fishing the river for more than thirty years. And they are both legally blind.

"With a name like *White Cane* everyone used to think we were drug runners," Randy tells me one day in the AGS cafeteria. "I love that. Blind drug lords disguising themselves as fishermen." Eventually his brother would retire and Randy would employ a trusted crewmate to stand at the helm while telling him all the visible landmarks, allowing Randy to continue to captain his boat.

To work in such extreme conditions, fishermen must have the modesty of Buddhist poets and the toughness of, well, fishermen. They must respect the laws of nature and accept the lay of the land or, in this case, the water. Fighting against the wild moods of the water, trying to outsmart the violent swings of the sea, will only get someone drowned. But quietly outmaneuvering its moods may get a fisherman more fish and ultimately a deeper sense of respect. Of course toughness and modesty are a rare mix with most fishermen.

What fishermen are reluctant to admit is that making mistakes is essentially the silent art of fishing. Everyone here makes them every day, but the good fishermen adjust and move on. I feel I am in the presence of a good fisherman when the mistake almost seems like a fun exercise in risk. Instead of showing fear he moves on so fast the mistake seems like an adventure.

So, fishing in Bristol Bay takes time to master. Reading the sea, the weather, and the tides requires years to understand. A fisherman may boast and speak loudly at the dinner table. He may even be obnoxious, but if he's been around long enough, there is a part of him that is quiet when he is on the water, always listening to the subtle clues nature is sending.

And when someone dies fishermen will speak of it only behind closed doors. They pursue the topic in the late hours of their drinking dens, where they discuss over and over why that person died. In that process they learn. Hardly any of this can be described as purposeful consciousness, a discipline often found in yogis, monks, ascetics, and people with more time on their hands than fishermen. Fishermen learn the hard way. By enduring one more season.

Yet wherever they come from, all fishermen have something in common: they believe they know how best to catch fish and how to save the troubled industry. Some say they should form a coalition to enhance their collective power. Others balk at this idea, saying that it's dog eat dog out there and each person is their own boss. With wild salmon from Bristol Bay going for fifty cents a pound, they say they are being exploited, used by the corporate giants with names like Trident, Peter Pan, Icicle, and Alaska General Seafoods. Usually these conversations end with heavy sighs, remembering the days when they used to get $2.20 a pound. They are being ripped off and they are going to do something about it. At least that's what they say.

"Fisherman" doesn't accurately describe what these people do. Yes, they make their money by harvesting fish; that is what goes on the dotted line in the IRS tax return. Some come with dreams of large money, and some come with high hopes to season their souls with righteous work. But the ones who stick around also have a jack-of-all-trades quality about them. To fish is to be part carpenter, plumber, welder, weatherman, cook, mechanic, businessman, mother, father, sister, and brother. Part priest, drill sergeant, cop, and outlaw. They have to believe in good and evil, and little prayers to the fish god don't hurt. They don't have to be great at any of these things, they just need to know how to make whatever's broken—crew, pipes, bed, wife, boat—work long enough to catch their fish.

Most fishermen are only fishermen part time. This includes the Natives. The work is seasonal, and there's not enough money in it anymore to last a whole year. Most of the fishermen who visit Egegik in the summer live in the Lower 48, and have jobs such as teachers, lawyers, housepainters, cooks, plumbers, warehouse managers, and artists. Some find ways to create luxuries in Egegik that remind them of their other, more civilized lives. Taylor Phillips, a part-time ski instructor from Idaho, and several other captains use a small shack outside the AGS bunkhouse, where they have a fully operational kitchen and dining table. Toward the end of the season they can be found there drinking wine from a box, eating thick steaks, and swapping stories of past seasons. Then there are the Italians from the San Francisco area, who have been coming here for almost five decades. They loudly express their disgust with how things are changing and use their hands to speak and expect special treatment from the fishery because they've been here so long.

The Orthodox Russians who fish in Egegik expect to be left alone and tend to speak to no one. Their children are easy to spot; they are the ones wearing head-to-toe clothing that resembles Mennonite garb. The male adults have full beards and their wives work in full-length dresses, usually with pants underneath. They never fish on Sunday, the Sabbath, and most live elsewhere in Alaska, in communities populated by other Russian descendants.

As for the crews, they come from all walks and stages of life. Many are sons and nephews and nieces trying to cash in on the family business. Others are students making extra cash for college. And some are professional fishing hoboes, drifting from village to village, boat to boat, season to season, working the Alaskan fishing jobs. Crew salaries vary, depending on their experience—15 to 20 percent of the take at the high end, and 5 percent at the low end.

Because of the stress on the body, working crew is understood to be a young man's job. Rarely will someone be over the age of forty, unless it's a family affair, where the husband, wife, and children all work together. Captains, on the other hand, are almost always over forty years old, and some stay at it well into their early seventies. There is no forced retirement for fishermen, and it's not a mystery why fishermen have a hard time quitting. Some can't quit because they make the bulk of their money from the summer job. Others can't quit because it has come to define who they are. Without fishing they are only defined by what they do the rest of the year: logger, tile worker, carpenter, or used-car salesman.

Nick Abalama Sr., a Native from Egegik, spent his last season fishing at the age of ninety-two being hoisted by crane every day from the dock onto the deck of his boat. Giving up the job meant much more than giving up a piece of income he had become used to. It meant giving up a lifestyle. In a world where rules are steadfast, fishing has always offered freedom. It allows a person to interact with nature in an extreme way and make a living doing so.

One more thing about fishermen: they are a greedy tribe. They are not likely to be seen at a Sierra Club meeting. They are more likely to be NRA card carriers and watch NASCAR on Sundays. They shop at Wal-Mart and most have a rough blue-collar edge to them. On the average they vote Republican even though every time I ask them which politicians they like they say they hate them all. Many think the locals are drunks and derelicts, and look upon Egegik as a doomed village that needs the fishermen to survive. This same logic allows fishermen to argue for hours on end that if they did not harvest the salmon there would be an overabundance of fish and the species would overpopulate the spawning grounds and die. This is the same logic that allows a big-game hunter to kill an elephant and then declare, loudly, that he was only "thinning the herd, keeping it healthy." The only logical response to this screwed-up thinking is to ask, "So how do you think all these salmon survived before we came along?"

My point? That without the constant eyes and ears of the law, and the hefty fines for infractions, fishermen would destroy the salmon run in a few years.

Yet there is something about the fisherman's mind that I envy. They don't often second-guess themselves. They don't doubt their right to reap nature's bounty. There is no moral dilemma. Instead there are salmon, food, and money. They are but deliverers of food, wanting to be paid justly for their service.

In fact that sentiment is yelled quite often over drinks at bars all over Bristol Bay.

"We feed the world. That's what we do."

I don't disagree. In fact, I believe they are correct. And every year after returning from Alaska I feel a surge of pride, something born from the act of working in such an ancient vocation. I too can be found in my local bar telling stories of how we fishermen feed the world. How I belong to a small union of people, as old as any profession on earth. How we help feed strangers in strange lands, people I will never meet.

Maybe one day the Alaskan fishing industry will disappear. Maybe global warming will heat the water to the point the fish move on. Or maybe it will be the new gold mine to be built on the shores of Iliamna Lake, where it will leak arsenic one spring and kill all the frylings. Or maybe the consumer market will shift because the Japanese, the most important market for wild salmon, begin to buy only farmed salmon.

Whatever the future holds it should come as no surprise to these fishermen. For the river teaches something about life. It teaches fishermen to be ready for change. They may not like it, they may resist with solidarity, but I doubt it. They are far too selfish and individualistic to ever truly join ranks. Besides, they believe down deep it's useless. Even this river beneath them, the one that gives them a bounty to live by, will surely change course. Change is inevitable. Finding a way to flow with it seems to be the key to survival.

After all, that's what the salmon have been doing for twenty million years.

MALIBU MARTY

THE FIRST TIME I MEET *MARTY* IS ON THE WATER, AT SHARON'S site. Dave jumps onto our boat, and immediately begins to tell us how he thinks the day will develop. He explains where he believes the fish are, what the drift boats are doing, and why the state troopers are idiots who don't do their job.

Then Dave turns to Marty, who stands in the front of Dave's skiff with a rope in his hands. The rope is the only connection from Sharon's skiff to Dave's.

"Don't let go of the rope."

Dave, used to working alone, broke down this year and hired Marty, a forty-year-old from California, as a crew member. Right now Marty has a hood over his head and sunglasses over his eyes, which don't disguise his resentment for having to stand in the skiff holding a rope.

For a few minutes Dave and Sharon share a cigarette and exchange family details concerning their father. Dave and Sharon have a tendency to get on each other's nerves, but they share a profound brother-sister bond.

Dave turns to me—he always tells his stories to the least experienced person in his audience—and explains the finer points of how to fish our net to its maximum capacity.

"Bow it out," he says to me, although he is really speaking to Sharon and Carl.

"Let the current work for you."

"Drag an anchor inside and close the gap."

Trapped and somewhat bored, I look beyond Dave and watch Marty, who is now staring at the river. Right in front of him a salmon fins to the surface of the water.

I see the moment Marty lets the rope go and lunges for the fish.

Also, from the corner of my eye I see Carl, who has also seen Marty let go of the rope. Carl says nothing and instead smiles and leans on the boat railing. When Dave starts in with fishing advice, Carl usually finds a way to get busy with chores or drift elsewhere in his brain. I always imagine during Dave's one-sided rants that Carl hears nothing but the birds above and the gentle slap of the water against the side of the skiff.

So, as Dave's boat slowly floats away I keep an eye on Marty, who is on his knees, his arms extended in the water, trying to catch the salmon with his hands like a blind kitten.

I turn to Dave, with the intention of telling him that his boat is floating away, but there is no hurry. There is no immediate danger. Besides, Dave hates being interrupted.

"Fucking AGS can't put limits on us. I am an independent contractor with a contractual agreement to deliver my fish in a timely manner. If they close the tenders then how do I deliver my fish in a timely manner? I can't. So they are making me break the law," says Dave. He is pontificating on the recent events whereby AGS limited each permit holder to a 5,000-pound delivery, usually imposed only when their capacity to process the fish at the canneries is overwhelmed with too many fish.

And Marty? He is now twenty feet upstream, drifting on a high-tide flood. Still focused on catching a salmon with his bare hands, he has no idea this is happening.

Finally, Dave turns and sees his boat, drifting upstream.

"What are you doing? You stupid idiot! I told you to hold the rope."

"Dude, I didn't know," says Marty, as he looks up from the water, his face twisted up in half a smile and half shame. His arms are soaked up to his biceps.

"You didn't know! What didn't you know? That the rope was the only thing holding the boats together?"

"No. I mean . . . I didn't know."

We pull anchor and rescue Marty. This will not be the last time we save him. And over the next three seasons Marty will oscillate between being an entertaining sideshow and an extremely tiring person to endure.

HIRING CREW IS a crapshoot and probably one of the most frustrating aspects of being a captain. Until a person is on the water, there's no telling how one will react. Over the years Sharon has been through dozens of crew members. There have been so many that she hardly remembers their names or where most of them were from. Most leave after one season, quickly realizing this is not the life for them.

Once she told me about a young man who woke up screaming every night, afraid he was drowning. He left one week into the season. Other times she's had no crew on the first day of the season and ended up hiring one of the drunks who drift in and out of Egegik during the season, looking for the odd job. Their commitments to being a crew member tend to run day to day, depending on whether they have enough money to buy a bottle.

Sharon tells me her first encounter with "Malibu" Marty, a moniker we used when he acted in a dangerous or clearly stupid fashion. She came out the front door and saw Marty driving Carl's ATV, with her son on the back. He drove into a steep hill and tried to bank the bike but almost flipped it. Sharon yelled for her son to get off the bike and then walked up to Marty.

"I told him to never, ever have my son in anything he's driving. Ever," she says. "Then he hit the gas and almost flipped the bike again. I told him he could never ride the ATV again. Ever."

His response to Sharon?

"I didn't know."

I lost count how many times he would walk into Sharon's cabin with his wet, muddy fish waders on, climb up the stairs and into Sharon's loft. All I ever heard, from my room downstairs, was Sharon yelling, "Get the fuck out of my house with your waders. Jesus! How many times do I have to tell you!?"

The thing with Marty is this: getting mad at him is easy. Staying mad at him is nearly impossible. He always smiles, takes the blame, and says he will pay better attention next time. And we always believe him.

Dave's patience with Marty begins to strain midway through their first season working together. One night Dave comes over late and is yelling about something to do with his table. I have ice on my arm and Carl is lying on the couch. We have an opening in three hours and are trying to get some rest.

"He carved his name in my table. Do you believe that? He took a knife and carved his name in my fucking table. Know what he said when I asked why he did that?"

There is a pause, although we both have the answer.

"I didn't know."

And there is more. Seems Marty needed a sponge to sop up a leak in his room. So he carved out a small piece of foam from the middle of Dave's bed, in the shape of a pillow. When Dave screamed at him, Marty held up the soaked piece of foam and said, "Look, it will go right back where it was."

But no matter how mad Sharon, Dave, Carl, or I ever get at Marty he never gives up. He puts up with the yelling and the disparaging remarks, and believes being a fisherman is a life for him. And every time, when I ask why he wants to keep coming back to a place that doesn't seem to want him, he answers, "I don't know. Maybe I'm stupid but this is the life for me. Being free from a regular job. And this nature. Can't see this shit in LA."

MARTY COMES BACK for a third season, but in control of his own skiff and site. He is now a captain of his own set net operation. He arrives a month early, which is almost unheard of in Egegik. He tells everyone he wants to be ready. In the off-season he has secured a temporary permit, by paying a $5,000 fee to someone who no longer fished in Egegik but still held the permit. Somehow he managed to borrow a skiff, and rents the site from Dave, who is charging Marty 30 percent of all his catch for the season. He has a used motor, which he bought for $5,800, a few dollars short of what a new one costs. And Marty signs the promissory note to pay for the motor before he ever tests the engine, which ends up not working.

I stop Marty in the middle of the road one afternoon, as he is hustling to camp for some spare parts.

"Why did you buy an engine that you hadn't tested in the water?"

He shrugs his shoulders and pulls his cap off his head in a show of anger.

"You can probably get your money back, right?"

"I don't think so," he says.

So, already deep in debt, he convinces Icicle, the fish company that has accepted him onto their books for the season, to front him a new motor on the books. They ship it from Seattle on a rush. This costs him another $6,000.

The season has not started and he is so broke that even if he has a great season he will still owe money to several people, not to mention that he won't be able to pay his one-man crew, a friend he convinced to come to Egegik for adventure and easy money. All of this is what awaits Marty, if he doesn't drown, which most people who know him believe is inevitable.

It only gets worse. A few nights before the first opening, around midnight, he decides to let off an entire crate of fireworks in Sharon's backyard. We have been building an extension off the back door, so there is wood and sawdust everywhere. Everyone in our cabin is asleep when the fireworks go off. I run upstairs to see what is happening and just then I see Marty dive into some bushes to hide. Sharon runs outside and confronts him.

"You are the stupidest motherfucker I have ever met. Don't ever come into my place again. Do you understand?"

Marty stands in the street, with Sharon screaming at him.

"Ah Sharon, I was just trying to have fun with you."

"Fun. You fucking idiot. Our house is asleep. There are no lights on. Did you see that?"

There is a pause and then from the window I see but can't hear the words being uttered.

"I didn't know."

And in this moment I am alerted to something frightening in Marty's character: he actually didn't know that fireworks would wake everyone up and perhaps burn down the cabin.

THE NEXT DAY his brand-new engine arrives. Marty spends the morning giddy with anticipation and ends up working several hours putting the engine together. I imagine he probably tells his crewmate that all their worries are over. That now they can catch some fish. Above, on the dock, some drifters ignore the set netters below, and instead pass their time talking and exchanging theories on how the season will unfold. Warren Hart is among them.

There are a few things anyone who has been in Egegik for a season knows. One is the ebb and flow of the river at the docks. It's a radical event that happens every twelve hours. Each time the tide goes out the sandbars are exposed, revealing the safe channel. After a few days even a rookie begins to remember the basic shape of the true river, visible only at low tide.

This is Marty's third year.

Having finished putting the engine on the skiff, Marty starts the engine. It purrs with power, giving its owner newfound hope. He looks up at the crowd, who by now are turning their heads and most likely talking about Marty's new engine. Ready for a test drive, Marty waves up to the veteran drift net captains as his crewmate pushes them off the dock. Then Marty accelerates, and drives straight across the channel and into the gravel bar. By all accounts the engine runs for six seconds before crashing the boat into land.

When Warren sees Sharon later that day his face is dead serious. "Stay away from that man. He is going to kill someone."

As Marty's season progresses it turns from farcical to dangerous. His net snaps on the first set, with no fish in sight. The next few days, whenever he throws out his first set, the net twists, taking hours to unwind. Weeks pass. One day he runs over to Sharon's house in his waders. He left his net in the water all day and the closing is in an hour. He missed the tide and his boat is now dry on the bank. He faces a large fine if he doesn't get his net out of the water by closing.

I am the only one home. I ask how he could let this happen.

"I was watching television and forgot," he says.

Reluctantly, I take the four-wheeler down to the river and push his boat back in the water.

As he shoves off he shakes his head and yells out, "Tell Sharon I'm stupid."

"I will," I say.

The last time I see Marty he and his crewmate are down at the company office to collect their home pack, which is the fish they have frozen over the season to take home. Home pack is arguably the reason why many people keep fishing, even when the prices are not worth the effort.

"Someone stole our home packs," said the crewmate, seething with disgust. It is the first time I've heard him speak all season.

"Not really," says Marty, with his aw-shucks attitude. "The guy acted like he did, but then gave them back to us."

"Still, I've been here two fucking months and didn't make a dime," injects the crewman, getting angrier with each word. "And now I have to deal with this bullshit. Being disrespected."

I feel sorry for Marty. He isn't a bad guy. He isn't mean-spirited or dishonest. But he still thinks acting innocent is an excuse for screwing up. Constantly. I have only been here one more year than he, but I know one thing for sure: he is not a fisherman.

"You coming back again?" I ask Marty.

"Hell no," yells his friend.

Marty laughs uncomfortably at his friend's anger, which is not amusing. He then shakes his head and smiles.

"I really don't know. Part of me says this place is evil, you know. Just sucks the life out of me. The people. Dave yelling at me all the time," he says, pausing in mid-sentence.

"But?" I ask, sensing his hesitation to end his life in Egegik.

"But I love it here. I think I can make it work. Maybe a couple more years."

No one knows how many fishermen have died in Egegik. Or Bristol Bay. I doubt anyone really wants to know. It's bad luck to know.

Yet every summer someone comes to Egegik thinking all there is to fishing is to buy a boat, buy the permit, hire a crew, and drop the nets. Simple, they are fishing. That person, who takes no time to learn the people or the water, will come home with no money or not come home at all.

I leave Marty and his crewmate and wander down to the river. Nearby, a local fisherman by the name of Billy Martin steps out of his shack with a bottle in hand and his sweatpants dangling dangerously below his ass. He is on his way to the liquor store, keenly aware of the one hour it is open in the evening.

I walk the bank, pass dozens of skiffs and drift boats being readied for winter. A tractor pulls one drift boat out of the water and drags it toward camp. There the boat will stay for the winter, balanced on wood planks placed on top of fifty-five-gallon drums. I keep walking, through dozens of seagulls camped under the dock, eating the scraps of salmon that empty through a chute from the huge grinding machines above. Farther on sit our skiffs, high and dry, at least until high tide returns. A tender is anchored where it always is, directly in front of Sharon's cabin, its cranes busy with fish coming in. I climb the bank, walk through the alders and grass, and reach flat ground next to Sharon's place.

Dave is in the yard reveling over a pile of nets he found at the town dump.

"They're brand new!" he yells as he wades through the webbing, like a child happy to have found a buried treasure.

Whose nets are they? Every so often the company goes through the lockers of fishermen who never return. They wait a few seasons, making sure the owners will not return. Then they cut the lock, empty all the contents into a truck, and drive them to the dump, all to make room for the next person who believes he will succeed as a fisherman in Egegik.

BUSH RADIO

I WAKE UP FROM A DREAM AND STUMBLE OUT THE FRONT DOOR. Night has overtaken the world, but there are no stars, only the Arctic wind blowing clouds down from the north. I stare straight ahead, trying to remember what day it is. The season is almost over. Soon clocks and phones and bills will reenter my life. But not yet, that's what I tell myself. I still have time to feel the blanket of nature wrapped around me like a warm fire.

A red fox runs by, never once looking over at me. I don't know how long I have been standing here. I am waiting for something more, maybe a wolf, maybe a moose or two? I find myself thinking I want to live here for a winter, when the wolves and moose rule the land. I will study them during the day and at night. Well, at night—that would be a problem. I am not sure I am equipped for winter in the Alaska bush.

One of the most perplexing aspects of traveling is this: wherever you go there are people there. Take the Sahara. In the middle of the largest desert on earth, surrounded by scorching heat and in an inferno of nothingness, a Bedouin will arrive on a camel, his face wrapped in a thick white turban. He will dismount that camel and offer a stranger a cup of water.

Or what of the Indians in the Amazon who have yet to make contact with outsiders? How do they live, adapt? And then there are the Eskimos who exist on the frozen permafrost above the Arctic Circle. Why do they continue to live there?

Closer to home, drive a lonely highway in America, round the corner of the last hill on the horizon, where desolation meets isolation. There will be a house, a trailer, a tent. Enter it and meet the people who live there. Ask them the whys and hows of living in a place like this and in the end the answer is always the same, no matter where you go in the world: "Because it's home."

It's easy to forget we humans can adapt to almost any conditions—eat any food and endure the harshest climates. Of course it helps if a person is born there.

Chuck, a local who spends many winters bouncing from the North Slope to Anchorage to Washington, tells me about the winter he spent in Egegik. A wolf lay down for him. That's what he says: "It laid down for me." That winter he trapped fourteen beavers and one wolf. The $3,000 he made selling the furs in Anchorage fed him for five months.

One afternoon he tells me the story.

"I was upriver and I saw it running across the ice up by Don's cabin. I wasn't sure what it was but took off after it on my snowmobile. I got closer and then realized it was a wolf. I stopped and took aim. Really wasn't sure I could have hit it, but it just kept running. Straight at me. Never slowed down. It would have run right into me. I've never seen anything like it. Heard about it from some people who used to hunt. It just laid down for me."

Natives talk about hunting like this. The wolf "laid down," or gave up its life, on purpose. It's not a question of debate or theory with Chuck, it's the way things happen.

This conversation takes place as we slow down on our ATVs. The motors idle as Chuck talks about winter in Egegik.

"I never want to stay here again for winter."

I ask him why, reminding him that he'd spent long, tough winters on the North Slope working the oil fields.

"It's lonely here. No one talks to you. And I can't sit inside like all of them and watch television. I have to be outside, hunting or trapping. And the trapping's not good anymore."

Just then Michelle, dressed in sweatpants and a short-sleeved shirt, comes up the road, pushing her baby in a carriage. A .357 pistol sticks out from the waistband of her pants. Mosquitoes swarm the child's face. Chuck and I part ways and I race down to the cabin just as Michelle goes inside.

"Why the gun?" I ask.

"Bear down by the town lake," she says.

I don't have the energy to reply. Instead I get a glass of water, my fourth of the day. Whether it's the work, the mosquitoes, or the booze, I'm always thirsty. We have been fishing for twelve days straight and have about six hours until we are at it again: setting nets, dropping anchors, picking fish, and then doing it all over again.

Yesterday Sharon asked if I am going to return next year.

"I don't think so," I said, feeling the pain in my hands and exhaustion of sleep deprivation. I had already decided not to return.

"I'll call you in the winter and we'll set it up," she said, laughing.

I am staring at the river out the back window of the shack. The river does this to everyone who stays here. Like a siren wanting us to watch its slow, languid movement, or a giant python sliding over the earth, it is mesmerizing by its very nature.

Soon Sharon is explaining to Michelle how to repair a net. I retreat to my room and lie down, not wanting to get involved in the current conversation about webbing. Instead I turn on my shortwave radio.

Whenever possible I listen to the open-air show, which broadcasts for a few hours each day. On this program locals send messages to each other using the bush radio. In the Alaskan outback, phones are a relatively new luxury item, and the radio has always been a more reliable way for locals to communicate with each other. I have spent many of my best hours in Alaska in Sharon's cabin, lying in bed, listening to the radio.

The only radio in the continental United States that resembles bush radio can be picked up in places like Farmington, Windrock, or Sells: reservation land, where the local Native Americans have a radio station that sets aside time each day for DJs speaking the chosen language. Other than that, the closest would be community radio in small towns when the DJ announces garage sales or reads off a list of happy birthdays. But even there, the DJ often brings an affected emotion to the announcement as if he or she feels the need to make the listener happy.

This doesn't happen in the open-air bush radio show. The DJ is only there to moderate the time spent on the air. The show is personal and the people who call in mince no words. Take this one: "Colleen, come get your moose." That's it. The voice is never raised in tone, not angry, not happy, nothing. And then the person hangs up. What's the name of the caller? No one knows, other than Colleen, who no doubt got the message and will soon be on her way to get her share of the moose. Where? That is between Colleen and the caller.

Many people use the show to wish each other happy birthday or to express their affection for a missed relative.

"Uh . . . okay. I want to tell my uncle happy birthday in Chignik and to tell Nora that I will not be there for dinner on Friday."

As I listen to the show, I always find myself imagining the size of Alaska. A third the size of the continental U.S., with a population of 600,000; or, put another way, 600,000 people living on 340 million acres, and almost half the population lives in the city of Anchorage. That works out to one person per 566 acres. For people not living in economic and political regions like Anchorage, Juneau, Fairbanks, or the Kenai Peninsula, where telephone lines are in place, making communication instantaneous, the radio is still a relevant medium. There are few roads to and from bush villages. Funerals, weddings, marriages, and announcements of children going off to college or to the army are made on the radio. For the most part people aren't speaking to the community at large. Instead, they are speaking to their extended family, spread out over a large geographical region. Even as satellite allows better cell phone reception, Natives still prefer the radio as their main source of information and communication. It represents their sense of community.

My favorite callers are the people who call in drunk and ramble on until the host reluctantly butts in and asks for the next caller. One time a caller had something to say about what he saw outside his window. He spoke in slow, stilted sentences, as only an indigenous person can—"It was really something . . . the eagle flew right over my house . . . and when I went outside there was one more. And the sun was setting."

Often, business gets transacted over the line—"Henry, I will trade the Yamaha for the freezer. Okay, call me or Joe." Or "Chuck, I need an outboard by Friday." No matter how many times I listen to the show it never ceases to amaze how a name is rarely offered by the caller.

The fact that the callers don't identify themselves doesn't imply they are being rude or forgetful. Just the opposite. It reflects the sense that they are speaking to family, to the people who know them by voice. The provincial nature of these people is served by the radio and vice versa. The familiar way in which they address each other on the radio is one of the clearest ways to understand the smallness of these villages, interlocked by blood and time. Everyone just assumes that everyone else is listening and if the intended listener is not listening they will get the message soon enough from someone who was.

Deaths are also announced via the radio. "George passed away last night. The funeral is Thursday."

The message is short and to the point, implying that whoever needs to know about George's death will find out and be at the funeral.

Of course the show is also used to arrange travel itineraries. Here is an exchange between the host and a typical caller.

"Hello, you're on the radio, go ahead."

"Hello?"

"Yes, you are on the air, go ahead."

"Oh, okay, uh hello."

"You are on the air, go ahead."

"Okay. I would like to tell my uncle to pick me up at the airport today, or tomorrow, when I get in."

The line goes dead for a few seconds and the host moves on to the next caller, a slightly inebriated person announcing, "I miss you, Grandma."

As for the caller asking his uncle to pick him up at the airport there is no name given, no village, no date, no flight time. But I would bet a season's wage that the uncle heard the message, knew it was his nephew, and was at the airport at the right time, sitting in his car and waiting for his relative.

THE FIRST TIME I ever heard bush radio was in Alaska, but more than a thousand miles from Egegik, in Barrow, the northernmost city in the state.

Barrow serves as the headquarters to the Arctic Slope Regional Corporation and is home to 4,500 people, mostly Eskimo, who derive from the Inupiat people who have inhabited the region for more than 1,500 years. The city lies 340 miles north of the Arctic Circle and in the dead of winter the sun doesn't shine for 51 days straight.

When I boarded the plane in Tucson, Arizona, it was a balmy 66 degrees. When I disembarked in Barrow it was 40 degrees. Below zero. That took some getting used to, but even stranger was that the cabdriver was from Thailand. Seems one man came over a decade before and his family, cousins, and friends all followed the money trail. Cab rides were $13, anywhere in town.

In my hotel room I flipped on the radio, my first experience of Alaskan bush radio. The host was speaking in the slow monotone I later encountered so often in Egegik. He was providing some weather tips for the citizens of Barrow.

"Remember . . . the oil on the skin will begin to freeze at thirty-two degrees below zero. That will be painful. So don't stay outside for more than a few minutes. Be careful and stay warm."

I heard the advice but my ridiculous curiosity got the best of me. I went outside, determined to take some photos of the sun as it popped up over the horizon, spraying the top of the world with two hours of golden light. It was a day I will never forget. The city ends where the Arctic ice pack begins. One doesn't forget the violent cracking of the ice; or the man on a snowmobile racing by, with a rifle strapped on his back and a dead caribou on a sled behind him. After being outside twenty minutes the pain began and I realized there was no one else outside. As I ran back to the hotel I noticed the small things, like an adult caribou frozen solid, standing in the yard of someone's home. Pieces of its hindquarters had been hacked off. People literally use the outside as a meat freezer. Holding my face in pain and embarrassment, I ran back to the hotel, past the front desk clerk, who smiled at my glowing red face. I had the early stages of frostbite.

From then on I listened more carefully to the bush radio.

Each summer after leaving Alaska I return to the Southwest and take a long drive into the hidden mountains of the deserts. Places with names like the Chiricahuas and the Gila Wilderness. They call these pockets of high altitude in the desert "sky islands." The area covered by these wildernesses pales next to the open space of Alaska, but why compare? Every square inch of preserved open space is something to be thankful for.

Driving out to these patches of the People's Land I often reflect on Teddy Roosevelt, the man who carried a big stick. What happened to the ilk of him, politicians unafraid to say they have a love for the land? Some politicians these days want to give it all away to the mining, lumber, and resource extraction companies. Don't get me wrong. Until we focus our efforts to harness the wind, sun, and alternative energy, we need all these industries. To deny it is a lie. But the scale is lost at times. Also, we tend to forget that left alone, the land works tirelessly for us. Day and night, every day of the year, the remaining open land of this country is working to give us air, food, and long stretches of nothing but a place to be reminded of where we came from.

THE MAIN REASON I take these long drives is to ease my return to a culture vastly different from the one I just left in Alaska. Also my liver usually needs a break. My mind is tired but my body taut. I tend to speak louder, a side effect of being around Sharon and her family and friends and the howling wind. To be honest, a part of me looks for confrontation, not necessarily a fight but someone to disagree with me about anything. I will yell louder and when they try to come over the top I will say I am a fucking fisherman. What are you?

So, I drive to slow down. I drive to allow my mind to drift and to remind myself not to forget the place I just returned from. Looking out over the desert my mind visualizes bears catching fish on the river, bald eagles stooped on a limb, and moose wading in a shallow pool. And the fish. I can almost hear them, by the millions, as they slap against the bottom of my car, swimming desperately upriver to complete their genetic mission.

On these trips in the desert I always turn on the radio, hoping to hear something, anything, that resembles bush radio. I try to imagine a person from the Lower 48 trusting the radio to communicate something as important as their arrival at an airport, or that someone has died. Of course it's impossible, not only because there are so few radio stations left in America that are manned by actual DJs but because the radio long ago lost its usefulness in our culture as a medium of import and communication. Instead the stations are manned by machines locked into programmed playlists and paid advertising.

However you look at it, no one is asking anyone to come get a piece of moose.

My own version of bush radio is playing in my mind when the phone rings the winter after my first season.

It's Sharon.

"See you in June?"

"Absolutely."

HOW WAS YOUR WINTER?

THERE ARE ONLY TWO SEASONS IN EGEGIK. WINTER AND SUMmer. It's either freezing or about to. Everything else is summer.

One summer, a few days after landing, a local Yup'ik man stops me in the middle of the main road to tell me about the death of Daniel Abalama.

"He woke up at six in the morning and thought it was six at night. He thought he was late for the liquor store and wanted to hurry before it closed. He jumped on his three-wheeler and went too fast around that curve—you know, the one by the bunkhouse."

I nod, knowing he is speaking about the only spot in town where one's view is blocked while driving. It is a place well-known for accidents.

"So he is going too fast and flips. The handlebar went right through his forehead. Killed him."

Strange, I thought, because I remember Sharon told me Danny's brother George also died in an ATV accident, a few years before.

The man shifts his weight from one foot to the other. He looks straight into the driving rain.

"How was your winter?" he asks.

I am trying to figure out a way to answer this seemingly simple question when I realize he doesn't really want an answer. He doesn't want to know details, he is just saying the equivalent of "How are you doing?"

"Good," I answer.

For a few minutes we stand in silence, staring at the flat world of the Alaskan tundra.

"Do you know the difference between summer and winter in Egegik?" asks the man, his head nodding back and forth.

"Ice?" I answer.

"In the summer we fuck and fish," says the man, his face now in a full smile.

"And in the winter?" I ask, playing the straight man.

"In the winter we fuck."

We chuckle a bit and then, once again, stand in silence. We do not know each other's names. I am in Egegik once again, where nothing is appropriate or normal.

"How was your winter?"

This is how the locals greet outsiders upon their return to Egegik. They don't ask about the spring, fall, or even the two months of the previous summer that weren't spent in Egegik. It's as if the eleven months spent away from here never exist: a person leaves Egegik, there is a pause in time, and then he returns.

If they don't remember you, then they don't ask you a thing or even make eye contact. If they do remember you, the conversation almost always begins mid-sentence, as if there has been no time lost.

"Where's Sharon?" a man asks me near the post office, the rain pouring down on us. He does not know my name, but I have something more important. An identity. I am part of Sharon's crew, thus I am accepted to a certain degree.

"Net locker," I say, zipping up my coat against the cold. In truth I never know where Sharon is, but I'm certain this man doesn't need the facts, only the outline of a picture.

"Tundra cotton is early this year," the man says after a few moments. His face is puffy from drink, his eyes black as coal. "Going to be a good year."

And with that the man turns and walks away, his pace slow and steady. I might see him once or twice more in the season, most likely at Sharon's during a late-night drinking binge, but we will never talk again, at least not until next June, when we bump into each other at the post office.

There is one person who greets me the same way each June: Kevin Deigh, Carl's brother-in-law. Kevin wears the same clothes every summer: Levi's pants, Levi's jacket, and a Marlboro baseball cap. He walks slowly, with a gait that looks painful. By late afternoon he will reek of rum, which he consumes at a level that would kill most people. His white tennis shoes barely fit over his feet, which are swollen with developing diabetes and gout.

Kevin runs the only store in town, which is owned by his father, who is the mayor, and his mother, who manages the PenAir in Egegik.

"Bill Tyler!" he yells.

I put a small jar of jam on the counter. It costs six dollars. Sharon and I once passed some time in the store looking at expiration dates. The oldest was a can of peanut butter stamped for use by 1992, eleven years past its due date.

"Bill Tyler. I thought you were dead," says Kevin, his grin so wide his eyes almost disappear.

Every season I look forward to seeing Kevin, even though he has a tendency to repeat himself, no matter the story or the audience. If he likes you he is full of charm and laughter. If he doesn't he has a tendency to ignore your existence.

Each year I tell him my name is not Bill Tyler and he laughs and says something about being a mountain man. I know Kevin likes guns and talking about hunting, but I also know he would rather leave an animal dead in the snow than have to skin it and bring it home.

So one year, toward the end of the season, after Kevin repeatedly calls me Bill Tyler and affirms I am not dead, I insist that he explain what the hell he is talking about.

"I will show you why I call you Bill Tyler."

We get on his four-wheeler and drive through the dim glow of sunset. It is just after midnight. I have a few beers in my jacket, which we open as we slouch in front of his television. Being in Kevin's house gives me pause. For a local, Kevin has it made. He has a job he can't get fired from, a vehicle with a constant full tank of gas, food on the table every night, and free airline access to anywhere in the United States, even though he hasn't been farther than King Salmon in almost three years. He also has an endless supply of rum, which most people agree will kill him one day.

The movie is titled *The Mountain Men*, and the story is about two beaver trappers stranded in the Rockies in the late 1800s. Charlton Heston and Brian Keith portray the trappers. As the movie plays Kevin drifts in and out of sleep. Every once in a while he opens his glassy eyes to say, "This is my favorite movie," and then drifts back to sleep. I watch, more out of boredom than interest. The party is already in progress at Sharon's loft and I need a break from the smoke and repetition of stories I have heard many times. Sitting with Kevin seems like a good idea, for now.

After twenty minutes of silence Kevin props himself up from his chair and points at the screen, his words barely audible.

"Here it is," he says.

In the film Charlton Heston enters the teepee of a Native American chief by the name of Iron Belly, which refers to the Spanish conquistador armor he wears. He's 110 years old and sleeping. Heston's character wakes the chief.

"Bill Tyler?" Iron Belly says.

"Yeah, it's me."

"Bill Tyler. I thought you were dead. . . . He who runs with the wind speak your heart."

Heston explains to the chief that they are running out of beaver to trap. He asks the chief where to find a new source of beaver.

"In the mountains of the river of wind, there is a valley. In the land of our enemies, the Blackfoot. Their hearts are bad, their eyes full of blood."

I turn to see Kevin. He's mesmerized. I swear he's trying to mouth the words, but then again I'm not sure if he's just mumbling to himself. Then he grabs the remote and plays the scene over again.

What does this movie, this scene, mean to Kevin? I have no idea. Maybe in his mind he is the wise old Native and I am the wandering outsider looking for riches? Or maybe Kevin wants to live in the fantasy of a movie that shows the fighting spirit of Native people in a time when animals were plentiful and Natives still had a free place in the world.

After watching the scene four times Kevin is hysterically happy and rambling about death and Charlton Heston. He yells, "Bill Tyler, I thought you were dead!"

Ten minutes later he is asleep for good and I turn off the television. I walk home to find eight people in Sharon's loft smoking weed, drinking, and laughing.

I put in my earplugs and climb into bed, determined to be ready for the next day of fishing.

Ultimately winters are what separate the people who live here from everyone else. If the cold isn't bad enough, there is the darkness and finally, like a crushing blow to the psyche, the isolation. Winter is the quiet killer in the Alaskan bush, when people snap. Violent stories leak out of Alaska from time to time, often blamed on what is called "going bush." That's what people say when someone kills people for no apparent reason, like the time a man shot seven innocent people in a village near Egegik.

An especially ugly example happened in McCarthy, a mining town six hours from Anchorage down a dirt road. One winter in the 1990s a man shot the mailman who had flown in the week's mail. Then the shooter hunkered down in the post office, the village social center, and killed the locals as they came for their mail. One by one. Almost half the town was killed before help arrived. I learned about the McCarthy killings from Malcolm Vance, who fished for many years in Egegik with his twin brother, Mark. He used to live part-time in McCarthy and was there the day several of his friends were gunned down.

Every June, upon my return to Egegik, someone has died over the winter, and hearing the stories of their deaths is like keeping score in some morbid diary.

"I DON'T LIKE all the outsiders coming here," says Wendy as we sit at the round table in the village office. Born and raised in Egegik, Wendy spends her winters, with her husband, working across the river as winter watchmen at an old cannery. She offers me coffee and cookies as we talk about how Egegik has changed over the years.

"Okay, we make a little money off the fish tax, but really they don't do anything for the town. They come, trash the place, take our fish, and leave."

We talk about her childhood.

"It's different now. With the alcohol and all. I mean every year we lose more and more people," she says, and smiles as another neighbor comes in for the free coffee.

I ask her to tell me what happened to Ken Chmiel, who went on a drinking binge over the winter, killing himself and another Egegik Native, a man named Enis.

She says it probably started with whiskey chased by a few six-packs of Budweiser, followed by more whiskey. This went on for days. That month the cold snap was especially harsh—temperatures hovered at minus 40 for almost two weeks. Ken shuffled in the dark from one home to another, taking the binge along like a long shadow dragging behind him.

She continues by telling me how Enis was falling victim to Ken's aggressive personality and drunken notions of great adventures. She and her husband tried to talk Enis back to their house, but Enis was already under the spell. Ken promised they would have the time of their lives, see wolves, maybe even shoot a moose.

They made it almost twenty miles upriver when Ken's truck went through the ice near an area called Don's Cabin, referring to a hunting shack on the side of the river, still owned by Don, the same man who owns the only liquor store in Egegik. Enis went down in the truck, probably dying very quickly. As for Ken, state troopers found him the next day inside Don's cabin. At 40 below zero he couldn't have lasted long. That he got to the cabin must have been an overwhelming struggle against the cold. The troopers found him fully clothed, frozen to death, with the smell of gasoline on some of his clothes. Nearby was a book of frozen matches.

I HAD MET Ken on several occasions but he never remembered me. The last time we spoke was toward the end of the prior season at Sharon's cabin. That was the night it became obvious Ken did not like outsiders.

Fifteen people were in the cabin that night, all in various states of inebriation. Kevin Deigh was slurring in the corner, tipping the rum bottle. His brother, Mike, was on the couch with Carl, laughing about something. Gabe from Club Ohio, was here as well.

Club Ohio is not the name of a boat. It is a consortium of people, not all from Ohio. Although the cast of characters changes over the years, the main players have been working Egegik sites for several years. They live on the far side of the river in a shack built of plywood, with an open floor plan. Inside the shack are four bunks, a kitchen, a wood-burning stove, and a wet room in front. Gabe, the loudest person from Club Ohio, was asking everyone in earshot for drugs.

"I will pay a hundred fifty bucks for a fucking joint," Gabe yelled at no one in particular. Everyone ignored him.

Except one local who said, "A hundred fifty?"

"Abso-fucking-lutely. I need a fucking fix. I'll smoke it, shoot it, snort it. Shit, I'll put it up my ass if it gets me high."

The local disappeared for fifteen minutes, returned, and the deal was done. Gabe smoked his $150 joint in ten minutes.

Meanwhile I poured myself a shot of whiskey and tried to hide the ice pack on my arm under my coat. A tendon had herniated in my forearm and my thumb was so useless that picking up a glass had become a challenging task.

By the time Ken entered, the room was full of laughter, which may be what set him off. He walked toward the back of the room, but not before Kevin Deigh started yelling about how much Ken owed the store. It was an old battle, carried on between the Deighs, who owned the store, and some of the locals, who owed them money—money the Deighs knew they would never collect. It was almost a forgotten subject, until people started drinking. Then all bets were off. Ken and Kevin stood face-to-face shouting about the overdue store bill.

"I didn't come here to fight about that shit," yelled Ken.

But Kevin was drunk, blind to his own repetitious tendencies. "You still owe. We carried you."

This went on for five minutes until finally something happened that I had never seen. Carl stood up and yelled, "Shut the fuck up. Kevin, you shut up about the fucking store, and Ken, just shut up. I don't want to hear any more bullshit tonight."

There was no more fighting.

That is, until an hour later when I walked up the stairs to speak with Sharon. At first I didn't see Ken. He was on the couch, out of sight. Before I could say anything to Sharon, Ken started in on me.

"Who the fuck are you?"

"I work with Sharon."

"I don't know you."

"He's okay, Ken. He's my crew," said Sharon, trying to placate him.

I told him we had met a few times, here in the cabin when he was looking for Sharon. He didn't react, his eyes were glazed with drink. I told him I'd worked for Sharon and Carl for three years.

"I don't know you. You don't belong up here."

He started at me, but I was not backing down. Not here, not in this house, where I had earned my place. Sensing something bad was about to happen, Sharon got my attention and gave me a nod, telling me I should go downstairs.

"Yeah, you greenhorn. Get out of here," Ken said, and took a long shot of whiskey. There would have been no fight because he would have beaten me silly in a matter of moments. Besides being drunk, he was bigger than me, much stronger, and most important he was angry. At what? No one ever knew.

That was the last time I ever saw Ken.

For years the U.S. government ran studies on the Inupiat in the most northern regions of Alaska, trying to discover one thing: how they endured so much physical pain. It was, and still is, believed by many scientists that these people possess something physiologically the rest of us don't that allows them to endure the harshest of conditions.

That winter, around the same time Ken went on a binge and drove into the ice, I was in Mexico surfing in an ocean warm as bathwater.

It may have been the first time I really pondered the question, "How was your winter?"

I'm still not sure how to answer this question, but I do know this: it's not the same as what's happening in Egegik.

I'VE BEEN HERE BEFORE

FROM MY DIARY, JUNE 22:

> No openings yet. The sun sets around 1 A.M. and rises around 3:30 A.M. Mostly everyone runs around trying to stay busy by fixing a bolt, tying a knot, sweeping the floor. The truth is everyone is waiting, just waiting.

A FEW DAYS later I am standing in front of the fillet table in Sharon's backyard, carving up a king salmon for dinner. Down below, on the river, I can see Sharon and Carl at the tender, four boats back in the line.

It's nine o'clock in the evening, and we have just finished nine hours of fishing. We bagged almost 10,000 pounds. The sun is blazing away, but the wind has died down to a whimper of a breeze, bringing the mosquitoes out in a feeding frenzy. I wonder how in a land so abundant with life mosquitoes are constantly short of blood. I quickly fillet the fish, throwing the head and carcass over the side of the bank and spraying small pieces of fish all over me as I swat at the buzzing creatures.

Once inside, I turn the oven on to 350 degrees and pour myself a shot of Jim Beam on ice. I lather the fillets with pepper and a bit of salt, spread some minced garlic and a small amount of Yoshida's, a sweet teriyaki sauce. Some people baste the salmon with brown sugar. Others use mayonnaise. I find that the less put on the fish the better it tastes.

My equilibrium is still off from the hours on the boat, and I sway back and forth as I wait for the oven to heat. The pain in my left forearm is worse than the day before. I convince myself to take comfort in this feeling—it confirms I used my body to its capacity today, and that confirmation soothes my inner voice, the one telling me I'm getting older, slowing down.

I open the oven and slide in the fish just as Kevin walks in. This is not Kevin Deigh, Carl's brother-in-law. This is Kevin Greggory from Holy Cross, a Deg Hit'an Athabascan village on the Yukon River. Shy, this Kevin stands six feet tall and has a head of black hair, a mustache, and a puffy face from his bouts with alcohol. He doesn't talk much, and when he does it's with a thick Native accent.

This summer he's been working with Earl, a white man from Montana who owns the cabin a hundred feet up the road from Sharon's.

"Hey," he says, very slowly.

"Just in time. Fish for dinner," I reply.

Kevin sits down and goes silent while I ramble on about the big day of fishing and how my arm hurts. He has not been on the water yet, something about having to mend nets and get an engine repaired. To change the subject I move on to the mosquitoes and talk about how nice it would be if the wind came back. Kevin nods, occasionally smiling in my direction. In two years I have spoken to Kevin only a few times. All I know for sure is that he got kicked out of his mother's home recently. When I ask him about the circumstances he doesn't answer. Instead he looks at his feet.

I offer him a beer, which he declines without uttering a word, shaking his head. Then after ten seconds he says, "Trying to be ready for tomorrow's opening."

I check on the salmon. A cream-colored glaze has covered the fish—fat oozing out of the salmon and mixing with the garlic. Fresh wild salmon baking in the oven makes me think of farmed salmon being served in restaurants all over the country tonight, platters of pale pink meat, artificially dyed. It has to be smothered with a heavy sauce because the kitchen doesn't want you to actually taste the fish. Hell, it has no taste. The fish in my oven, on the other hand, is a deep crimson and smells fantastic.

And now a few more words about farmed fish. Upon hatching, farmed fish are immediately fed hormone-enriched food. They live in mass schools in a pen, stuck in the water somewhere off the coast of Chile or Norway. Unlike their cousins swimming about in the Pacific, eating a diet that has nourished their species for thousands of years, farmed fish dine on a mixture of synthetic food, antibiotics, and their own feces.

Also, salmon naturally produce a slimy layer that lathers their bodies in a protective coat. This slime is what makes them so hard to grab from the water. Wild salmon maintain a low level of slime because of the fact that they are darting around in the ocean water. Not so for farmed salmon, who end up eating and breathing the slime from the thousands of other fish in their pen.

In short, either I eat wild salmon or I don't eat it at all.

I AM TALKING about farmed salmon to Kevin, when he suddenly speaks.

"Weren't you in South Naknek in 1990?" he asks.

I don't know how long I stand silent. The shock resonates in every cell in my body.

"You worked in the ovens."

"You were there?" I finally ask.

"Yeah, worked on the line. It sucked. . . . I remember you were dating that really cute girl."

I don't move. I don't talk. I just stare back at Kevin, who is smiling, happy that we have a memory in common.

Kevin is correct. In 1990 I spent six weeks working in a cannery in South Naknek, about ten miles north of King Salmon and about fifty miles east of Egegik. I was twenty-four and had just completed a two-year trip around the world, mostly living out of a backpack. Two years outside the country, and not much had changed. Still, more people watched the Super Bowl than voted; budgets for defense outweighed education, medicine, and everything else combined, except for our national debt, which seemed to be an invisible chain around our collective ankles.

Broke, and having a hard time reentering American culture, I figured the best way to ease the process was to be in America but as far away as I could get from its hyperactive culture of strip malls and sitcoms. So, in June of 1990 I landed in King Salmon and took a small plane across the Naknek River to South Naknek. My object was to do what I've always done when I feel out of place—work jobs that exhaust my body, giving my mind time to catch up with the world around me.

Years later, when Sharon called and asked me to work in Egegik, I didn't know where the exact town was, but I knew it would be close to Naknek, a place to which I had vowed never to return.

What Kevin doesn't know, can't possibly know, is that I followed that cute girl back to California and we ended up living together. We were young and in love and it didn't take long before we planned a life together. We were in the final stages of getting ready to take a trip to Mexico when she fell asleep at the wheel, hit the back of a farm truck, and died. My life changed radically after the accident. I wandered for almost two years, taking odd jobs, bumming in the West

Indies and traveling South America. But no matter how far I traveled or how drunk I got with the locals the memories always found a way of seeping to the surface of my mind. Finally, and for reasons I can't fully explain, I ended up going to Sarajevo, at the height of the Bosnian War. It was madness, it was bloody, and it was the best thing I ever did. I met people there who cured me of my own grief.

How many days have I walked past Kevin without saying a word? More than I can guess. There he is, standing under a swarm of mosquitoes, working on Earl's nets. Sometimes I race by on the four-wheeler waving in his direction, but only out of courtesy or desire to make contact with another person. He always nods back but rarely does he ever say a word, until this day in the kitchen.

Kevin barely travels outside western Alaska. Most of his adult life he's migrated from places like Egegik to Naknek, to Holy Cross to Chignik and places in between. He knows very little about the Lower 48 and even less about the rest of the world. Every season he shows up late and walks into Sharon's house with the same look—something between disappointment to be here, again, and relaxation, as if he's resigned to the path life has chosen for him. And each year he bubbles with excitement over his recent finds of music CDs and DVDs.

"I got Def Leppard's world tour on DVD. Journey too. And I have Cher and Bon Jovi, brand new."

Sharon and I always thank him for his offers and tell him we don't listen to those bands. He nods and smiles, probably happy he can keep them for himself.

That he remembers my face, from another time and space, is what causes me to pause in the kitchen.

I have no desire to bring up my own past on the Alaskan Peninsula. But there is something important in Kevin's memory and his timing of sharing this memory. It tells me something about the Native mind. They love to tell a story in stages; as if time is simply a nuisance they haven't yet taken seriously. Sometimes the conversation consists of one word. "Fish," a man says, his smile widening. Always I am left slightly frustrated, wondering if we have the same definition of "fish." When speaking, they often delay their response to a question, maybe ten seconds, a minute, or an hour. The effect of the delay can make them seem thoughtful and pensive. That may be true of some, but over the years I've come to realize that they just have a different sense of time. Or, said another way, Native Americans, unlike the people of more modern cultures, don't believe talking is the same thing as thinking.

After saying a few words, or nothing at all, they may just get up and leave. This might even be considered a good visit. To further confuse the matter, the following day the same person or persons might be jacked up on whiskey and cocaine and talk through the night, laughing and screaming about nothing in particular.

Sometimes I think of the girl I met here ten years ago. It was definitely on my mind that first season when I realized how close I was coming to my past. But after a few days on the river my nerves calmed and I was consumed by the work. By the second season it got better. By the third, I thought less of the girl from my past and more about the people I had met in Egegik and how isolated my life was during this one month a year from the rest of the world.

Eventually, the memory evolved into something else entirely and it gave me great hope, as if I was participating in something ancient: recycling ourselves to live again.

SOURDOUGH

IN THE DARKNESS I WALK TOWARD CAMP FOR A LATE-NIGHT shower. I have two cans of Budweiser hidden in my pants pockets. My Carhartt jacket fends off the cold wind, leaving my face to take the brunt of the sideways rain.

The hot shower has been one of my personal sanctuaries in Egegik. After a few days of fishing, the mind and body begin to hover in a liquid state, somewhere between jet lag and utter exhaustion. Sometimes it feels as if a hot shower is the only way to prevent passing out.

After a day of fishing, some people eat dinner, relax, and drift off to sleep. Not me. Within ten minutes of getting out of my waders and wrist guards I am walking to the shower. I grab a towel and clean clothes, put them in my backpack, and walk the half mile. Rain, wind, or hunger can't keep me from taking a shower. The stink from the fish and the dried sweat from the work are unbearable.

There are two showers, both small stalls, with just enough room for a modest-size human to stand in but not to bend over. In the span of twenty-four hours, dozens and dozens of fishermen will use these two tiny showers.

Still swaying from twenty hours on the boat, I try to stand still and let the hot water pour over my aching body. I put the beers in the soap dish and finish them both by the time I turn off the water.

Next, I bundle up and walk straight home.

One afternoon on my way to the showers I see Kevin's boss, Earl, standing outside his home.

Five feet six, with a hunch in his posture, Earl likes to squat near the ground when he speaks. Like a bird crouched low to the earth, this is where Earl does his best thinking.

Like many people who have spent a lot of years living in Egegik full-time, Earl doesn't talk much. Known as a loner and a quiet man, Earl rarely invites anyone inside his home. This particular day I walk past and wave. He waves back. I pause long enough to see he is tinkering with an engine.

"What you got there?" I ask, knowing Earl tinkers as a way of life.

"Differential burned out, I think, but if I can find a Yamaha I can probably fix it."

As I ponder a reply, I notice Earl's cigarette teeters on the edge of his lips as he pushes and pulls and bangs around the piece of machinery.

"How are your birds?" I ask.

"Four ready to hatch any minute," he says, looking up in a hurry and smiling through his thick gray beard. His eyes are still hidden by his baseball cap.

"Excellent," I say, already feeling as if I have overextended my welcome. Turning into the wind, I sling my backpack over my shoulder and head for the shower.

"Want to see 'em?" he yells after me.

Like Carl, Sharon, and Dave, Earl Goodman first came to Egegik, almost twenty years ago, to work as a crew member for Sharon's father, Warren. In the early years Earl took his earnings from the summer and went back to Montana for the winter to hunt and trap. But eventually he built a house in Egegik and began collecting various objects to fill it. Over time the yard filled up with old boat parts, pieces of metal, rusted tools, rope, damaged nets, broken-down engines, and almost anything else that ever passed through Egegik. As if the weight of all these objects wasn't enough to keep him in Egegik forever, he took it one step further: he mounted a television high on one wall, put a reclining chair directly in front of it, and went about his lifelong passion of breeding cockatiels.

The birds perch in a series of cages lining an entire wall. The heater is set to the low eighties, making the room almost unbearable.

"Gotta keep it constant for hatching," he says, sitting down in his recliner, the only chair in the room.

Next to the chair is a five-pound coffee can full of cigarette butts. Several butts lie on the floor, and there are more on dishes nearby. He smokes three packs a day, with a seamless transition from one cigarette to the next.

I've seen Earl on his boat getting anchors and nets ready with an ash more than an inch long, just sitting there, as if reluctant to depart from its owner, waiting for Earl to give it a little shove. Then the moment the cigarette is done, he flicks it away with one finger while the other hand grabs the next one. His fingers are stained yellow, as is the patch of beard near his mouth. If he didn't actually have to sleep he could probably knock off four packs. And during the peak of the fishing season, when twenty-four-hour days are normal, he may even be able to begin a fifth pack.

When first coming into Earl's house it is clear Earl only spends time in the room with the chair, the birds, and the television. There are two other rooms, but entering them would be impossible. Stacks of clothes five feet tall ooze out of the doorways in a suspended state of falling forward, like a frozen cascade of water.

"Don't get in there much anymore," says Earl, taking a drag and watching my eyes take in his home.

And although I stand only a few feet from the kitchen, I can't see it. Like a hidden clue in a clever child's game, the kitchen only begins to take shape as the eye adjusts to the overwhelming piles of newspapers, open cans of food, boxes, dirty dishes, tools, clothes, and scattered pieces of paper that cloak the stove, sink, and refrigerator. A pot of beans occupies the only available space on the stove. From the pot, a spoon stands straight up, most likely there for days, if not weeks. When I ask Sharon and Carl about Earl they aren't sure what he eats. They can only testify that he has a strict diet of cigarettes and coffee.

I ask Earl about a video game that people in town say keeps him inside all winter.

"This damn game has got me all tied up," he says, as he grabs a computer off the top of a broken VCR and a pile of tools. He quickly makes clicks on the keyboard, allowing him to enter the computer game. As far as I can tell it's a fantasy game that takes place somewhere between the Middle Ages and Middle Earth. He shows me knights and dragons and horses and box canyons that lead to mysterious caves.

As he tells me about the game I can't help but let my eyes drift to the small trail of bird shit that extends from the cages to the rug and is making its way toward his chair, like a slow-moving mold. Near the doorway to the rooms in the back is a half-eaten ear of corn. As I listen to Earl continue to explain the computer game I realize something important about his life: he sleeps in the reclining chair.

"Yeah, a couple years ago just decided it was easier to sleep right here in the chair," he says, confirming my suspicion.

In the winter Earl looks after Don's liquor store while Don suntans in Hawaii, flush from all the money he earns selling booze and tobacco to the fishermen and cannery workers during the season. As for the rest of the year, a steady trickle of local alcoholics is enough to keep the store in business.

For his winter duty Earl earns no money. Instead he settles for free room and board and a supply of free cigarettes. Only recently has Don agreed to give Earl a cut of every carton he sells.

"I don't sleep in the bed there either. Don's got a reclining chair," replies Earl when I ask about Don's accommodations.

• • •

Earl is what people in Alaska refer to as a sourdough, or an old-timer. The term's origins date back to the Klondike gold rush when settlers did not have yeast as a starter to make bread. Instead they used sugar, water, and flour to collect yeast from the air, and thus became known as "sourdoughs." Today many people use the word *sourdough* to mean someone who came to live off the land or to live the Alaskan dream, and in that process turned "sour" on the land, but doesn't have the "dough" to leave.

"Thinking of selling and moving back to Montana," says Earl, lighting another cigarette.

"Want to leave but fishing just ain't paying the bills lately," he explains while taking one of the cockatiels out of a cage and placing it on his shoulder. The tropical heat in the room is making me sweat.

"I don't want to be here. Don't really have any friends and I'm getting too old to fight against the aggressive fishermen."

I mention the high prices of living in Alaska. The fuel. Food. Flights to Anchorage. I tell him that anywhere he lives in the Lower 48 will be cheaper than living in Egegik, that all he has to do is get out. He stares at the birds and then back to the television, craning his neck upward from his chair. For a moment he reminds me of a burned-out patient in a county hospital room.

"Yeah, I guess so. If I sell the permit I could get enough together to leave," he says without emotion. Then he stamps out his ash and lights another cigarette.

Upon leaving, I take the long way to the showers. I round the post office and walk along a row of houses. I don't know who lives in all these homes; there are never faces in the windows. Children don't play in the yards. I look in all directions, but there is no one there. I pass a house with a Harley-Davidson sitting on the porch. What's that doing here, I ask myself. Overhead, the wind vibrates against the wires, creating an eerie noise that, at the time, sounds precisely like emptiness.

THE FOOD CHAIN

FROM THE MOMENT A PERSON STEPS OFF THE PLANE IN EGEGIK a sensation forms in the back of the mind. It can also be felt on the skin, like the soft itch of a sudden rash. Those who stay only a day or two shrug it off as geographical disorientation. They tell themselves it is a case of culture shock or something to do with the wind, which can drive a person crazy in a desperate search for shelter.

Those who have spent time in remote Alaska know this feeling as something else entirely. It's called *Ursus arctos horribilis*, also known as a grizzly bear.

City dwellers aren't familiar with this feeling. They know fear as a defensive emotion specifically related to urban phenomena: fear of carjackers, of being robbed in their own home, or of being the victim of a random crime.

But they do not fear being eaten alive.

There is no other feeling like it, not even in war. The only thing I feel when entering a war is a heightened sense of caution. Surviving war takes luck and some savvy calculation, like how to avoid a drugged-out soldier who only wants the sunglasses off my face. This kind of fear has a quality of despair and has more to do with facing the craziness of man.

But knowing you could be lunch? That's something else entirely.

ALTHOUGH ALASKA is home to some of the largest animals on earth, most animals here are harmless, at least to humans.

Some people have a fear of wolves, but that fear is entirely baseless, a result of Hollywood's making horror movies where wolves turn into rabid monsters with glowing eyes. In truth, in a hundred years there have been only a few attacks by wolves on humans. Yes, they are known to prey upon young cows and sheep; something the ranching community in the Lower 48 likes to advertise as a reason to shoot them.

The caribou are harmless, unless perhaps they run you over in a stampede. Once Carl and I went upriver and saw a dozen caribou sloshing around in the mud, trying to get to shore. They did, but for a moment there I felt kinship with the caribou, stumbling over nature's obstacles.

A single salmon can't hurt you, but they can turn deadly once they reach critical mass on the deck of a boat. Many a boat has been lost because the weight of a catch has taken it down in rough weather. So, it seems the fish do occasionally get their revenge.

Then there are the lynx, wolverine, and coyote, all animals that a visitor will rarely see in Alaska. If they do, the encounter will be fleeting as the animal runs for cover. The beluga whale, the whale shark, and seals are all much too busy feasting on salmon and other oceanic delights to give humans much notice. The same is true of the killer whale, which spends its time chasing all the seals, whale calves, and fish.

Strangely enough, the deadliest animal in Alaska is the one people most associate with a comical figure: the gangly yet ferocious moose. Most injuries and deaths attributed to the horse-sized moose are a result of car accidents. But moose can also move surprisingly fast when threatened, explaining why people are sometimes trampled to death.

Still, it is the grizzly that owns the dark places in our minds, where fear multiplies and festers with every detail learned of their attacks. They don't attack often, and the attack can usually be traced to something a human did wrong. But when they do attack in earnest, survival is rare.

Grizzly bears are actually broken down into two subspecies. The first is the inland type, known simply as the grizzly. This big bear lives in the mountains and inland hinterlands of Alaska, Canada, and a thin wilderness belt between Canada and Yellowstone National Park. Its cousin, known as the coastal grizzly—or Kodiak, if standing on Kodiak Island—is the one that lives in Egegik and all other places in Alaska where salmon spawn. These bears, which stand up to ten feet tall and weigh up to 1,200 pounds, are bigger than their inland cousins—anywhere between 200 to 400 pounds heavier and a few feet taller—due to the single fact that they gorge on highly fatty salmon. Coastal grizzlies eat up to 20,000 calories a day when the run is strong.

MY FIRST CLOSE encounter with a coastal grizzly comes midway through my second season. Around 1:30 in the morning I wake, needing to relieve myself. We have been fishing steadily for five days and thus my mind is already in a state of sleep deprivation. I stumble out of bed and open the curtain to see a dim glow of twilight. I take a deep breath and turn toward the back door. The door has a deadbolt but no doorknob, meaning there is nothing to keep the door shut, except if the deadbolt is fastened.

Then I hear it. First the breathing: a loud and heavy sniffing on the other side of the door. I stare at the deadbolt. It is locked.

Next I hear a loud bang against the door, which bends the wood inward. I freeze in place, unsure how to react. I turn my head and look, again, out the side window. This time I see a giant mother grizz rise up out of the tundra with one of her cubs at her feet, playing. It is not a spring cub, more likely two years old and weighing between two hundred and three hundred pounds. The noise at the back door is the other cub trying to get inside.

After hearing me, they all run off, leaving me standing in the window in a full sweat although the temperature in the cabin hovers around 40 degrees.

I decide to pee in a bucket.

Here is the scary thought. The back door on Sharon's cabin tends to close quickly after opened, the result of a heavy door and imperfect carpentry. If the deadbolt had not been in the lock position, and the cub had entered, the door would have quickly shut behind him, trapping the young bear inside. It would have panicked and at that point our lives would have changed forever. The mother bear would have easily torn through the thin plywood wall. An attack would have been inevitable. Yet, later, when the events that led to the attack were studied, it would have been obvious to most Alaskans that the attack was a result of human error; for having a door without a handle.

ONE AFTERNOON I follow some bear tracks toward American Creek, upriver from Sharon's place. I keep one eye on the thickets in anticipation of a grizzly jumping out to rip me apart. The other eye stares intently at the shoreline, hoping to find washed-up artifacts from the past. But it's no use. Nothing lasts here. The ice crushes everything in the winter and in the spring when the ice melts, mud covers everything with thick sludge, hiding even the largest of objects. By summer the tides swallow all of it whole, every twelve hours pulling anything loose out to sea. Yet occasionally these tidal shifts accomplish something else. They uncover objects that have been buried by a thousand winter storms and several feet of mud. Sometimes the Natives find mammoth bones buried in the ground since the last ice age. The best beachcombing is out on Goose Point, a thin strip of sand separating Egegik from the pounding waves of the Bering Sea. There people find ivory in the form of walrus tusks. They also find car engines, washed-up pieces of boat, refrigerators, batteries, guns, and anything else man carries out to sea.

But the most common artifacts are glass fishing floats, an early-twentieth-century version of corks, most likely made by Japanese and Russian fishermen. Just like today's corks, which are made from Styrofoam, the glass corks, some the size of basketballs, were used to float the nets on top of the water. Long before plastic and nylon, ropes were wrapped around the glass corks like a spiderweb to secure them to a net. Glass corks can be found all along the shores of Alaska, but the prize ones have Japanese writing on them. No one I ever ask knows what the Japanese characters mean, but I have always imagined the markings are from sake bottles melted down by the fishermen on the passage over from the motherland to the rich waters of Bristol Bay.

Walking along the beach I don't find anything better than a red Styrofoam cork. Then, finally, I spot something else—the sixteen-foot carcass of a beluga whale. We noticed it last night on our way across the river to set our nets. Its head had been chopped off. Either some locals wanted to cook up a pot of beluga brain stew, or it was shot by a fisherman who quickly got around to cutting off the injured area to hide the evidence of a crime.

If ever there was a creature made to travel these waters it is the beluga whale. The first time I came in contact with one of these magnificent creatures I heard it long before I spotted it. The time was just after sunrise and a thick fog hovered ten feet from the top of the river. Our nets were out, and we were quietly sitting in the skiffs waiting for them to begin working. The sound was unmistakable: a creature blowing out a loud whooshing gush of air. I scanned the horizon but saw nothing. My eyes played tricks on me for a few minutes, until finally I saw a large disk of white slicing through the water. I've seen whales before, mostly humpbacks cruising the California coastline. But these were different. There were at least fifty of them and they had no large dorsal fin, just a smooth body of white that appeared almost liquid as they bent through the water, like a pale white wheel cutting through steel. Their grace took our breath away and we said nothing as the gang of whales cruised downriver chasing salmon.

I couldn't help but think of the many times I've listened to the fishermen gathered around the docks, sipping their coffee, plotting their revolt against the whales.

"A whole mess of belugas eating up all the fish. We lose our whole season to the fucking things."

"Gotta shoot 'em or something."

"I'd get rid of all of them."

I WONDER WHEN exactly did we become so arrogant as to believe we are the most important species on this planet? When did we lose the imagination that allows for the possibility that we are not the rulers of this world, but merely one more creature trying to make its way? This arrogance allows us to assume the salmon will keep coming by the millions, for our consumption, each and every year. And if they don't return, that same arrogance will lead some to say it's not our fault—because of overfishing, oil spills, toxic runoff from a mine, or global warming—but instead the whimsical ways of nature.

One summer I met this arrogance in the form of a dentist when I spent five weeks across the river at a sporting lodge, where Carl worked. The dentist couldn't stop talking about shooting a grizzly bear. Or a wolf, a moose, or anything else that moved in the bushes. One night at dinner he talked loudly and endlessly about how he killed wolves and coyotes.

"Got my snowmobile an overhaul to make it go faster and I just run them down. One right after another," he said, with a large grin on his face.

As for bears, he was even more fanatic. One night he set out several salmon carcasses in hopes of getting a bear to come up next to his bedroom window. He couldn't shoot it, as they were not in season yet, but that wouldn't stop him from the pleasure he got out of acting out the motions of shooting it.

A bear came and the next day at breakfast he told the story over and over—how he would have shot the bear, with what gun and what caliber of bullet.

And we let these people walk free. We let this one in particular fill cavities in our mouths. I wonder if he weren't a dentist, confined by the tedium of the job and the addiction to the money he must make, would he be a sadistic serial killer? I spent many free moments during his visit imagining him being torn limb from limb by a bear that he failed to kill in time. The thought seemed cruel and yet justifiable.

Carl and I have spent a great deal of time arguing the merits of hunting bear and moose. I argue for killing something only if it's going to kill you, or if the meat is part of your regular diet. The rich, predominantly white men who come here don't care that much about the meat, only the hunt and the trophy racks.

Carl also finds trophy shooting to be pointless, but working as a guide helps pay his bills, not easy to do in the Alaskan bush. He's never shot a grizzly and has no plans to, and as for moose, the Alaskan hunting regulations only let the hunters shoot old males, leaving the young to grow the herd. After they shoot a moose Carl takes the body carcass—which the hunters rarely want—to Egegik, where he spreads the prized meat among various locals. And who can argue with that? Moose has been their main source of food throughout the long winters for thousands of years.

For the most part the hunters come heavily armed, overdressed, and underschooled in the art of being in nature. Instead they are focused on the art of obtaining. Obtaining one more prize for their wall, which usually has other exotic animals—gazelles from Africa, cougars from Wyoming, and maybe even a poached lion's head. And they come to Alaska for the grizzly.

Being around something as majestic as a grizzly bear makes a person wonder about life itself. And about God. What God? I'm still not sure what to call God—whether it's a force of nature, or a guy with a beard, a sword, or a fat belly—but I am pretty certain that God doesn't want a rich guy from Texas with too much money hanging the heads of the earth's great creatures on his wall. No matter the God one prays to, that must be near the top of the sin chart.

People speak of grizzly bears as savage. They talk about the bears' natural state of violence, as if they are on the loose, lying in wait to maul any person who crosses their path. But the facts don't tell that story. The truth is they are relatively easy to avoid. They have their natural ranges but are not territorial. They take lots of naps and go to great lengths to avoid people. But, like humans, they will defend their young by any means necessary.

So, as I stand on the bank of the river, where thoughts drift in and out like tidal pulses, I wonder who is more savage, the dentist or the grizzly? I fear the rabid dentist, with his turbo snowmobiles, high-powered rifles, and total disregard for life, more than I fear all the bears of Alaska put together.

HIKING IN SOUTHERN Arizona I once met a man who had recently shot a mountain lion. I was hiking near Warsaw Canyon, not far from the Mexican border. It's in a remote range of mountains, a natural corridor for animals from north to south and back again.

The narrow dirt track was carved high above a deep canyon, so when the man saw me he brought his truck to a stop. I stopped walking but kept my guard up. This was smuggling country. Strange things happened on this road.

He came around the back of the truck, where I was standing, and he opened the tailgate. It slammed down and the man let out a loud holler. A whoop. In the back of his truck lay the freshly killed mountain lion, shot through the neck.

"He was just sitting there on a ledge not doing a damn thing. I took aim and fucking nailed him."

The man looked over at me with a smile as wide as his head, but

I said nothing. He was wearing a sidearm and obviously had a rifle nearby. Even if I were to knock him unconscious he could wake up and shoot me farther down the road. It was either take him out or walk away.

I walked away, not saying a word. After all, what do you say to someone like that? I can only hope he overindulges at his local bar and eventually just fades away.

I am often asked why I keep returning to Egegik. The money is not great. The weather is brutal and the work is both difficult and dangerous. And at the end of each season I promise myself I will never do it again.

I return to Egegik because I need a place where nature still has the upper hand, reminding me that my existence is fragile and fleeting.

As I walk upriver, I am here and now. I take great comfort in places like this. People hunt to eat. People fish to eat. People laugh because they can't help it. People drink to forget the harshness of where they live. And people dress to stay warm. There is no fashion, only function and dysfunction. The trappings of the world disappear, unless a dentist flies in and ruins it for me.

Out here there are no roads or hotels or sporting venues, just the steady and relentless howling of the Arctic wind. Many find these kinds of places inhospitable, barren, and boring to the eye, but the truth is we need places like Egegik and the mysterious lands hidden beyond its horizons. These are the places we call the wild, the last places where nature and man can travel backward thousands of years with the single chirp of a bird, or by hearing the roar of a grizzly. These places reconnect us to our primal senses and to nature's theme song. If wilderness is to survive us, it will be in places like the Alaskan Peninsula, where the salmon run forty million strong and humans still break out in a sweat because the grizzly is at the top of the food chain.

THE LAST FRONTIER

On a layover to King Salmon, I find myself in a bar in Anchorage sitting next to a man dressed in jeans and a button-down shirt. He is young and well-spoken, and his mannerisms display a man who is at the top of his game, which turns out to be prospecting.

"People think with today's technology and wealth that minerals are found by satellite, or geeks reading infrared data. Mining companies have all that, but before they make a decision to invest in a place, they send people like me to the location to prospect, just like they did a hundred years ago."

He goes on to tell me that he spends most of his life being dropped in remote locations in the world with supplies, a shotgun, a few local assistants, and a satellite phone. Once on the ground he looks for colors of rock, shapes of mountains, geologic clues to what lies beneath.

"Gold, copper, zinc, diamonds—all of them have signatures on earth. I spend a lot of time panning streams looking for clues."

He tells me the story of the two men who, after spending their life savings and burning through their marriages, finally discovered a massive diamond field in Canada. The story is intriguing. The major diamond companies were also looking, with more money and more manpower. Everyone was searching the same area, located near a series of lakes, but no one could find a thing. Then finally one of the men sold off the last of his assets and he and his partner flew to Canada, certain they had the answer. They did. Everyone had been looking and testing the ground surrounding the lakes. The men decided to look *under* the lakes.

"They sold out for tens of millions," said the man with a large grin. "Wish I would have found that one."

I ask him his success rate, as a prospector.

"If I was a baseball player I would have been fired a long time ago," he answers, laughing. "But companies don't always pay me to find something. They also pay me to find nothing."

THE LAST FRONTIER. That's what it says on the Alaska license plate. For the tourist the tagline conjures up visions of glaciers shimmering in a radiant blue light, and mountain goats standing against the wind on unexplored peaks. People come to fish on the Kenai Peninsula, where anglers can be found in July standing elbow to elbow catching decaying red sockeye. They come by cruise ship, motor home, and plane, all to have a piece of what is the idea of Alaska. They want to see an untamed land, the place where Jack London got drunk and dreamed up stories of the Far North.

But for the oil, gas, and mining businesses the last frontier has always been a constant cash cow. To big business, Alaska has always meant one thing: extracting resources from the land.

First it was the Russians, whose fur traders took what they could and then sold the frigid land to the United States in 1867 for $7.2 million, or two cents per acre—arguably the best real estate deal in history. Then the Klondike Gold hit, bringing thousands over the Chilkoot Passage heading north, in search of easy money. Of course the timber industry has been hacking around the Alaskan woods as long as white men have been here. But these were all small pockets of investments, well hidden from the sight of the Alaskan pioneer. That all changed in 1956, when big oil was struck on the Kenai Peninsula. Congress, until then, was content on letting Alaska be a territory, meaning it couldn't be taxed and regulated like a state. This arrangement favored big businesses and their ability to do what they wanted and how they wanted, without paying Uncle Sam. But once the black gold started gushing, Congress decided it was time to declare Alaska the forty-ninth state of the union and collect taxes.

Since its statehood in 1959 not much has changed. Alaska is still almost two and a half times the size of Texas and governed like a fiefdom, controlled equally by political and corporate interests.

IRONICALLY WHAT MAY eventually kill the Bristol Bay salmon run may not be fishermen but a Canadian company named Northern Dynasty Minerals, which is proposing to dig what promises to be one of the largest mines on earth: the Pebble Mine.

If built, the mine would be an open pit located near the north end of Iliamna Lake and would measure two miles long, a mile and a half wide, and seventeen hundred feet deep. The take from the mine has a projected value of forty-two million ounces of gold and twenty-four billion pounds of copper. And three billion tons of waste.

Iliamna Lake sits one hundred miles to the northeast of Egegik and covers a surface approximately the size of Connecticut—eighty miles long, twenty-five miles wide, with a surface area of more than a thousand miles, making it the largest lake in Alaska. Extremely remote, the lake has a mythology that dates back to the original tribes that lived on its shores. They believed a large animal up to twenty feet long lived in the water and to this day people spend time and money trying to find the creature.

It's no secret that open-pit mines are the worst offenders when it comes to toxic releases into watersheds. Arsenic, cyanide, and mercury are the tools of mining, and even under the safest conditions they trickle into the land that surrounds them. It's an unavoidable consequence of digging a pit half a mile into the earth's skin.

Historically, Iliamna Lake is also the spawning grounds of one of the largest runs of salmon in all Bristol Bay, and therefore the world. But for the past decade the Kvichak River, which flows from Iliamna, has been a protected run, meaning fishing is either extremely limited or not allowed at all. The reason for this protection has to do with a noticeable drop in salmon returning to spawn. The biologists who study salmon behavior are unsure how to explain the sudden drop in the Kvichak salmon. It could be the weather, species straying, or possibly it was overfished for too long, not allowing enough salmon to spawn.

But if the mine is allowed to be built, there won't be any more debate on why the salmon went away. Overfishing will be the least of the worries facing salmon.

I'VE ALREADY SAID that salmon are resilient creatures. They come home to their natal streams, lakes, and rivers to spawn, but they are also survivors, meaning if their home stream is altered somehow, in a way they don't like or recognize, they will move to more hospitable water. Fish biologists call this straying and in the past this has happened in times of cataclysmic earth changes. The most recent example of mass straying by salmon happened after the volcanic explosion of Mount St. Helens in Washington on May 18, 1980. The volcanic eruption, besides destroying forests and animals for dozens of square miles, also created deadly mudflows, which boiled and buried all the fish in the north fork of the Toutle River. The only salmon of that generation to survive were strays that did not return that year. Likewise, the recent repopulation of the Toutle River is also partially due to strays coming back many years after the cataclysmic event.

Salmon are an indicator species, a term used when the health of a species acts as an indication of the environment around it. More plainly spoken, if the salmon disappear we are in serious trouble. The salmon runs in Alaska today were not here ten thousand years ago. Before that the area was glacial, meaning there were no running rivers. No one knows where the original salmon runs came from, but what is known is that they came here once the ecosystem could sustain their cycle of living. And if they leave, that will also be a telling sign.

When I ask local people about the Pebble Mine some believe it should never be allowed to open. Others believe it will be a good thing, bringing jobs and money to a severely impoverished area. And others, well, they don't want the mine, but they want the money.

It seems clear that mines like Pebble are the enemies of a fragile ecosystem, but I am not surprised by the acceptance of the mining operation by the locals, both Native and others. This is not the first time I've heard these conflicting ideals. They want their natural world to remain intact and yet they also want the money. Years ago, on that winter trip to Barrow, I got into a conversation with an elder Eskimo about hunting rituals for whales. He explained the ancient rite of taking the whales and how food is shared among the villagers. He spoke with great passion about their rituals, even producing a drum, on which he pounded, that is used to call the whales to the shore in order to give themselves to the Natives. He said he was gravely worried how global warming was melting the ice cap, something that he feared would one day threaten the whales and the Eskimo way of life.

Then I asked about opening up the Arctic National Wildlife Refuge (ANWR) to the oil companies.

"Yes, that would be good," the man said, without a moment of hesitation.

But, I pointed out, drilling for oil is harmful to the land. Their land. And that oil causes the global warming that is melting the ice.

He nodded his head and smiled, as if he recognized the irony of the situation.

"Yes, but we need the money today. I can't worry so much about tomorrow."

• • •

AS IF THE salmon runs in Bristol Bay aren't already imperiled, big oil is beginning to lobby Congress to open up the bay for offshore drilling. This has been a long-standing fight between the oil companies and the people of Alaska. The only thing stopping oil drilling in Bristol Bay is a word that sends shivers up and down most Alaskans' spines: *Valdez*.

On March 24, 1989, the oil tanker *Exxon Valdez* struck a reef, spilling a total of eleven million gallons of crude oil into the Prince William Sound, forever altering the waters and surrounding landscape. Today this environmental disaster still lingers in the Alaskan psyche, and the damage done to the ocean and wildlife is still being felt. The fish, bird, and mammal populations are still suffering, and towns like Cordova, whose main source of money was fishing, may never recover from the hit to their local economy. But, as drilling for oil in Bristol Bay becomes more and more likely, the locals can be heard once again fighting for both sides of the battle. They want the money that comes with leasing out their land, but they want to preserve the salmon, a species that has defined their culture for almost eight thousand years.

As I ponder what will happen to Bristol Bay in the coming years, I remind myself of Carl's take on the Pebble Mine. He put it quite simply.

"It's gold," he said, shaking his head, in Sharon's loft one night. We were halfway into the bottle and staring into the river and the flat tundra beyond. "I mean, what the fuck good is gold? You can't eat it. You can't burn it. And it might kill the only thing in this part of the world that feeds people. Salmon."

Then I asked about oil in Bristol Bay.

"Ten years. That's what I give this place before some outside assholes kill it with their newest way to make quick money."

ONE THING IS certain: the delicacy of this ecosystem is difficult to imagine. In the more temperate zones, flux and change in the landscape are more readily absorbed by the ecosystems. In many ways the harsh landscape of the north is much more fragile than that of the continental United States, or the rain forests in the equator region. In the north, even small changes in the climate can kill all that lives here.

If the rain isn't just so, or if the wind blows in the wrong direction, the fish don't swim upriver. The moose don't come down from the mountains if the temperature in the summer isn't just right. Caribou can disappear for no known reason. And if the fish don't return the grizzlies and wolves don't return. The eagles stop showing up and the river goes fallow, eventually driving away the ancient cultures that have lived off the salmon for thousands of years.

Oil companies and their lobbyists and politicians don't take this into account when using words such as *footprint* or *exploratory*, modern-day euphemisms for *rape*, *pillage*, and *excavate*. And these men and women who sign the papers and push the buttons and transfer the money sit in high-rise office buildings in Washington DC, New York, and Houston, far removed from the places they are about to destroy.

IN THE BAR in Anchorage I order another beer and ask the prospector about the future of Alaska, in terms of the natural resources.

"Well, it's rich in resources and there is not much political resistance to mining. So it's perfect."

"But what about protecting some of those resources?" I ask. "Isn't saving the natural state of Alaska also a worthy goal?"

"Yeah, well, that's one way to look at it," he says, unfazed, taking another sip of beer and another bite of food, his eyes locked on a sporting event on the bar television.

The writer and naturalist Edward Abbey, in his day, might have gone to the site of the Pebble Mine and sabotaged a few tractors and burned down the double-wide office buildings on the site. Today, I'm not sure that would work. Today it will take an act of great civil courage. It will take voters in the Lower 48 to demand Alaska not be used as a playground for big business. It will take the will of the Alaskans, all of them, not just the Native people or the city liberals but a majority of Alaskans, to proclaim that the salmon are worth more than gold.

THE MAYOR

Every day Dick Deigh wakes up, drives down from his house on the hill to his other house in the center of town with his wife at his side, grabs a cup of coffee, and watches *The Price Is Right.* Every day. Unless he's fishing, in which case he occasionally misses the show. But that's where to find the mayor in the morning. At his table, with his back to the large window, watching lovely models hand out washing machines, trips to Greece, cars, and bundles of cash to perfect strangers. For reasons I can't fully explain, I've watched snippets of this show in Africa, where men argue over the correct price of a blender, made in China. And in Hong Kong, I've sat on a small bench staring through the glass of an electronics shop, watching the show with several strangers, all arguing over the price of a mattress. All over the world, the show plays and the people who watch seem to be hoping they could be in the shoes of the contestants, but for just a moment in time. They would make the correct decision, guess the right price, and win the prize. But Dick doesn't want to be on the show. He has no need for the products. For Dick it's purely entertainment.

"Goddamn idiot. He's a fucking moron guessing that price," he yells at no one in particular. Everyone else in the room ignores him. In the room, on any given day there are grandchildren; his wife, Scovi; Jennelle, his daughter, and two sons, Mike and Kevin; Carl, who is his son-in-law; and his brother, Bobby Deigh, who runs heavy equipment for the village. On most mornings Bobby stops by for coffee and to watch the show or, as he likes to say, watch Dick watch the show. Dick yells, he screams, but they all know better. He's not mad, not even close. How can he be? Dick Deigh has it made.

Dick and Scovi own two homes, this one near the cannery, where he can be found during the day, and the other one near the back of town on a hill surrounded by 117 acres, his Native allotment of land. It's not fancy, but definitely a step above the other homes in Egegik. In the winter Dick and Scovi go on cruises and make a few trips to Las Vegas to gamble.

"We always stay at the MGM Grand. And every time we go to the restaurant they remember us. Imagine that. Of all the people that come through their doors every day and they remember us."

I mention that the job of casino employees in Vegas is to remember people, but Dick ignores me and turns his stare back to the television where Mr. Bob Barker is giving away a *brand-new car*.

Dick is seventy-three years old and stands almost six feet tall, a bit more with his baseball cap, which always sits on the very top of his head, threatening to fall off any moment. His eyes are sharp and alert, never revealing a hint of being tired or distracted. He's always paying attention, even when he's yelling at the television. He has a rough way about him, maybe falsely so because he's losing his hearing and tends to yell instead of talk. Want some propane? "Come tomorrow. I'm busy now!" he might yell as he moves from the barn-sized garage to the house. If I bring up the name of any person in town he gives a one-line response that will range from "Good people" to "Idiot," and occasionally "Dumbshit."

The most important part of Dick's character is that he laughs loudly. He embodies the type of person it takes to live in a place as isolated as Egegik, where family politics and interpersonal drama rule the day. And he has the temperament to deal with the summer influx of so many strangers to his land.

It helps that Dick owns the only store in town, open a few hours a day. The prices can astound even the most hardened Alaskan bush person. Chili runs six dollars a can. Cereal boxes seven dollars. A small bag of grapes? Eight dollars. There used to be store credit, better known as PAF: payable after the fishing season, but that ran out a few years back.

"I have a hundred fifty thousand dollars on the books, still owed to me by people in town. I'll never collect it," yells Dick one day when I ask what happened to the credit system.

Some say Dick likes it this way, people owing him. It allows him to have markers on people in case he ever needs them. Maybe, but you wouldn't know from the way he talks about it.

"Degenerates. All of 'em. It's PAF around here, but some of these bums just don't want to work."

One day sitting at the breakfast table with his brother, Bobby, they speak about a time when everything worked on a cash basis, when people gambled the titles to their boats, their season's earnings, and sometimes their wives in backroom poker games. Then they remember further back, when the fishing boats were wooden with a single sail. Because there were no motors, each Monday of each week in June and July the sailboats were dragged behind a schooner called the monkey boat out to the bay.

"That's how it used to be. On Monday the monkey boat would haul us out, seven boats tied up to each other by a rope. Everyone was out Monday through Friday. Never got off the boat. Had all our food, a bucket to go to the bathroom. For cover from the rain we had a canvas tarp. It was a different time. It was hard, but we did it."

This is how the fishery was worked until 1952, the first year gasoline-powered engines were allowed on fishing boats in Bristol Bay. By the 1960s the thirty-two-foot law was introduced, not allowing any commercial salmon boat to be over that length. It was meant to discourage large vessels from sweeping the fish up in large quantities. As the decades passed the boats just kept getting fatter, deeper, stronger, and lighter. But never longer than thirty-two feet.

ONE AFTERNOON toward the end of the season Dick calls Sharon's house. He needs a favor. His crew left early and he needs a few hands to work the deck of his drift boat, named the *Naughty Lady*. His younger son Kevin used to crew for him, but this summer he gave up fishing to run the family store. Dick's older son, Mike, usually fishes with the mayor for two or three weeks every year, but this year he had to leave early and has already left for the season, back to his home in Washington.

So, Dick needs crew.

"What am I going to say? It's Dick Deigh," says Sharon when she tells me she volunteered herself and me for the job. Carl is also recruited, but he has no choice. Family relations obligate him.

The next day, fifteen minutes into the opener, we are stuck on a sandbar, the boat completely out of the water, tilted 45 degrees on its keel, right in front of the town dock. The mayor is in jeans and tennis shoes sitting behind the captain's wheel. Carl and his son, Griffin, are lying in one of the beds down below, taking naps. Sharon sits next to me and we are peppering Dick with questions while the rest of the fleet cruises by in the deep channel, no more than fifty feet to the east.

"Goddamn it. You guys were distracting me. I will never hear the end of this shit. I don't get stuck on sandbars."

Dick was born in 1933 to a father of Scandinavian descent, which accounts for his pale skin and blue eyes. His mother was 100 percent Aleut, which gave him his Native face, and probably his warped sense of humor. I've seen a picture of Dick in his navy uniform and his movie-star good looks are still there, his eyes wide and charming.

From where we sit, waiting for the flood of high tide to get us off the mud, we can see the house he was born in, across the river from the village. His brother, Bobby, was born in a house just below that. Dick enlisted in the navy at age eighteen and spent four years based in Long Beach, California. When he returned to Egegik he brought with him his first wife, who he says was a beautiful blonde.

"Then the damndest thing happened," he says while recounting the early days of his life. "She started sleeping with my cousin."

I ask what happened after that.

"Well, she eventually left."

And the cousin?

"A couple of years later he was found facedown in the mud. Dead," he says, his face melancholy. "I think it was all a shame."

Dick looks out to the water, checking on the boats downriver moving back and forth, looking like small models in a kid's bathtub.

"Goddamn. I haven't been stuck on a sandbar in a long time."

Dick was the first fisherman to take a drift boat upriver and throw out sets, using the sandbars as end points, effectively trapping any fish going upriver. His rough-and-ready reputation of being a smart captain had its benefits. When the boat builders delivered his first drift boat, it contained a surprise for his crew: lacquered in polyurethane to the bottom of the fish bins was a poster of *Playboy*'s Miss June 1968.

"I tell you what, when you are fourteen and you know there's a naked woman under the pile of fish, you pitch those bastards a lot quicker," says his older son, Mike, when I ask him about the pinup girl.

For a while other fishermen assumed Dick was fishing upriver to avoid the harsh weather that plagues those that fish out in the bay. But soon they realized Dick was catching fish. Lots of fish.

"Biggest year was two hundred twenty-six thousand dollars. In cash," he says. That's a quarter of a million dollars in four weeks of fishing. "Couldn't happen today 'cause the price is so shit, but then it was two twenty-five a pound."

I ask about the drugs that poured into Egegik during the big seasons.

"I tried that cocaine shit, didn't do nothing for me. None of that crap did. Well, weed was okay, but goddamn if it didn't make me hungry," he says, laughing. I spot Carl down below; his eyes are open and he is quietly laughing along with Dick, Sharon, and me.

To pass the time, Sharon takes Carl's son out onto the mud, which is hard packed. They walk out several hundred feet before turning around. Griffin seems giddy that he is walking on the bottom of the river. On deck Dick curses the boats as they pass by, but I know he doesn't mean it. Most of them are his friends, others his relatives. Every once in a while the radio crackles with someone calling the *Naughty Lady*.

"Dick, you catching anything?" asks Ron Hart, Sharon's half brother, who has fished in Egegik for more than thirty years. He owns the *Ocean Magic*, a drift boat, which he captains while his wife, Lynnie, and two daughters work on deck as the crew.

"Oh not yet, just waiting for the flood," answers Dick, half smiling.

"Good idea. Hate to have you be in a hurry or anything," Ron says, with a trace of friendly sarcasm. Dick pushes his hat back to scratch his balding head. During the season several fishermen in Dick's posse will use code to tell each other where the fish are, and aren't. The code is used to throw off other groups of fishermen that are listening and using their own code. These groups of fishermen don't share their profits, as Sharon and Carl do, but they share their information. And friendship. Today, everyone knows Dick is caught up on a sandbar in the main channel, but no one will dare say it in plain language on the radio. That would embarrass Dick, and they won't do that.

IT'S HARD TO sit with the severe leaning angle of the boat, so Dick puts his hip against the seat and looks out the side window. It's sunny and the wind has died down, a beautiful day.

I ask him what was the stupidest thing he ever did when he had the big money.

"Nineteen sixty-four," he yells. "It was a big season and I was in Naknek. The poker game went on all night and by the time it was over I won eighteen thousand dollars and the title to a brand-new Ford Thunderbird. I gave the car away and spent the money."

I ask why that was the stupidest thing he ever did with his earnings.

"I could have been the one that lost that night and that would have been stupid. Only stupid people gamble with their money. I never played another game of poker."

When Dick isn't fishing he is usually busy running errands for PenAir, which Scovi manages out of the front room in their house in town. She coordinates all the pickups and departures from surrounding landing strips, usually populated by fish camps and a few locals. A busy destination during the summer is International Seafoods of Alaska (ISA), a large set net operation located downriver, facing the bay. Planes land on the beach during low tide. If it's high tide they land on a small airstrip in the tundra called Coffee Point. There is also an airstrip behind Bartlett's Lodge, directly across from Egegik, where Carl sometimes works as a fishing and hunting guide.

If Dick is the mayor of Egegik then Scovi is the traffic cop of the village. She wields her power to get a person in and out of Egegik with an iron fist. Standing just over five feet tall, she has the knack to intimidate almost every human being who ever sets foot in the village. Scovi yells even louder than Dick, but unlike Dick she rarely laughs, at least during the busy season. Knocking on her door takes a great deal of bravery, or complete naïveté.

Some mornings I tiptoe around the house, peeking in the side window to see if Scovi is in the room. If not I will slip in and sit at the table to listen to Dick talk.

One morning three contestants bid on a lawn mower. They've all obviously overbid, something Dick can't stand. But even worse, the last contestant bids higher than all the other three, effectively taking himself out of the race. The smart money is to bet below the other three, picking up the win by letting the other three be disqualified by overbidding. "They must watch the show before coming on but still they get on and become fucking idiots," yells Dick as he turns his head to look out his window.

"That moron," he says of a man passing by the window. "He kisses everyone's ass."

The radio comes to life. Someone at ISA is calling for Scovi.

She appears from a back room and yells into the radio. The woman on the other end gives details of a crewman who needs to get picked up today.

"On the beach?" asks the woman.

"He can't get picked up on the beach," yells Scovi. "Tell him to go to Coffee Point."

"But he's already on the beach," the woman says, hesitantly.

"The tide is too high! Go to Coffee Point!" she snaps, and shakes her head.

ON THE BOAT, stuck on the sandbar, I ask Dick about his wife.

"I don't know why she yells at everyone," Dick says while laughing, "but I love it."

It's not hard to imagine that if Scovi had the same job anywhere in the Lower 48 she would be fired in her first week. But this is the Alaskan bush, and having someone competent and reliable here to do the job is half the battle. Besides, she is very good at what she does, just not good to the people she is supposedly hired to help, the customers.

"I grew her," says Dick, smiling at me. I must look confused by the word *grew* because he continues. "I was twenty-eight and she was fourteen and she just kept chasing me until I gave in." Sharon looks over and I try to imagine how it must have been in Egegik forty-five years ago, when they met.

"She was the ugliest person I had ever seen," he says, laughing, which makes us all laugh.

"But goddamn, she is the best thing that ever happened to me. And I love her. I don't care how mean she is to fishermen. Most of them are all fucking idiots anyway."

I wonder what it would be like to live an entire lifetime in a place like this. What genetic tools are required, what mental stamina has to be sharpened? This thought keeps me curious about the locals and at the same time makes it impossible to judge them. They live here, I don't. Maybe a case of beer a night is what it takes for some. Others develop a hard exterior personality to keep people from harassing them in the dead of winter. Still, I can't put it all together. But maybe I don't need to. Egegik is a place where racial slurs are accepted, pedophilia is condemned but rampant, and alcoholism is the 800-pound elephant in the room. In this context Dick is a champion, seventy-three years of hitting the ball out of the park.

FINALLY THE TIDE rises and we lift off the sandbar like a giant cork and head out to the bay. But we are too late. The flood started over an hour ago, the best time to fish. So the rest of the day is worthless as far as fishing is concerned, and the only excitement comes when Dick scares off a seal by shooting his .22-caliber rifle in its direction.

By the time we drop anchor it's almost dark and we have 1,400 pounds, enough to pay for gas and pay me and Sharon $150 each. There are no exchanges of thanks or invitations to the house. Dick gets in his truck and shuts the door.

"Gotta get me a few more Poon Tangs for next year," he says.

Sharon and I laugh as Dick drives up the beach to his house, a hundred yards away.

I first learn of what he meant by Poon Tang a week before, sitting at Dick and Scovi's table. Outside it is dark and raining, the wind blowing around 30 miles per hour. There is an opening in two hours. I have to get back to Sharon's house soon and get ready to fish.

The door opens and in walks a small Vietnamese man.

"Come in, Poon Tang," says Dick. "Come in."

The man is part of Dick's crew. He puts his hands in the air, as if looking for direction.

"What time we go fishing?" he says with a Vietnamese accent.

"We go soon, Poon Tang," says Dick.

"I fix net. And fix more if you want," says the man.

"I'll pay extra for that, Poon Tang. Thank you. I'll come get you when we leave."

The man smiles, nods his head, and shuts the door behind him. I wait a few seconds and then turn to Dick.

"Poor bastard needs money," says Dick. "He's a shrimp fisherman in Louisiana and hurricanes have wiped him out, totally. Got a wife and kids."

"Okay, I give up. Why is that man's name Poon Tang?"

Dick smiles. His brother, Bobby, laughs and nods his head. Scovi walks through the room.

"Assholes," she says, and keeps on walking.

"A friend told me he was a good worker," says Dick, smiling at Scovi. "So I hired him. He's a great worker. No bullshit. Knows how to mend nets, set line. He's a fisherman through and through, not like these pussies that come and go every year up here." At this point I am almost sure I blush, feeling as if he is pointing his harsh words at me.

"Anyhow he told me his name on the first day and it's some crazy Vietnamese name. So I said, no way. Pick a regular name—Bob, John, Tom—I don't care, but pick something I can pronounce. Three days go by and finally I ask him, do you have a name yet? He says he can't think of one. So I say, okay, from now on you are Poon Tang."

By the end of summer Poon Tang is the most popular crewman in all of Bristol Bay.

LATER I THINK about all the outsiders who have come and gone in this village. Most never meet Dick or Scovi. I am lucky because I work for Carl and Sharon, people who have deep ties to this community. This is not a place to make friends on the cheap. It is a long process, one earned over many seasons, by hard work, loyalty, and keeping one's shit together when things go bad.

The truth is that Dick carries a big bark but not much bite. He is not a grouch. Neither is Scovi. They are two people in love who found a way to survive in a place that, by its very nature, waits for people to die early. Or go broke. Or get lost in the bottle. But Dick and Scovi remain steadfast in their commitment to protect each other. The yelling? The brash interaction with outsiders? It's not an act. It's a shield.

"Fishing today with Dick?" asks Scovi one day as I stop by looking for Carl. Sharon and I are trying to fix an engine and she sent me to look for Carl to help us.

"Not today," I say, always ready to flinch from this small woman's wrath.

"Poophead," she says, with the slightest trace of affection. Seeing an opening I turn and speak.

"Scovi? Can I ask how you and Dick met?"

"I'm not telling you anything," she says, now smiling, and walking toward the back door.

Her smile disappears as the door slams behind her. I stand for a few moments in her living room and watch through the window as she hangs the laundry, Dick's clothes.

It is almost 9:00 A.M. *The Price Is Right* is on in a few minutes.

TOO MANY FISH

MOST OF THE OLDER NATIVES, AND HISTORICAL FISH CHARTS, will argue that the peak happens on or near July 4. This massive wave of fish can last anywhere from one day to three or four days, depending on the tides, the number of fish, and the weather conditions. Everyone has a theory, but fish are known to move in large pulses when prevailing winds push them upriver on big tides. Often this wind comes with a punishing rain.

The peak is what captains plan for all season. In these few days their share of the haul allows them to pay for most of their overhead expenses for the season, which lasts three to four weeks. Basically, fishermen can't survive financially if they miss the peak. This leads to a heightened anxiety felt throughout camp in the buildup to the peak. The gossip grows louder and often takes on an air of worry as it bounces from boat to boat like sparks of electricity.

"The fish are down the peninsula but are supposed to be coming this way."

"It's over. They already went to Naknek."

"We need more rain. Fish love the rain."

The only other thing about the peak worth mentioning is that threat of injury or death increases dramatically during this time.

Even with all the worry and anticipation, sometimes the peak catches everyone off guard. For us, this happens on June 26, 2005, during my third season.

The day starts with a strange omen. For two days the wind has been howling through camp with gusts to 50 miles per hour. The rain has come down in alternating waves of drizzle and downpour. But by the time we set out in the morning the wind has disappeared, there's not even a breeze. The sky is dark, but it's strangely balmy out. We pick our nets twice in one hour, giving us half a tote. Then, completely unexpectedly, the temperature drops drastically and a bolt of lightning strikes the water.

"These are aluminum boats, right?" I ask Carl, who, like the rest of us, is transfixed by the lightning show.

"Yeah," he says, shouting through the rain and looking back at Sharon.

Immediately Sharon starts the motor and drives full speed for shore.

We drop anchor and hustle up the bank and walk straight out into the tundra. We lie flat on our backs on the soft, squishy ground and stare at the violent but beautiful sky, dark to the north and sunny to the south. Then the hail begins, pelting us with ice balls the size of large marbles. I pull my raincoat over my head and turn to lie face-down in the tundra, the hail bouncing off my back side.

"Ever seen anything like this?" I yell at Carl over the noise of the hail.

"I've never even heard of anything like this," he yells.

• • •

We stay on the tundra for half an hour, letting the storm blow over us, around us, and through us. Finally the sun comes out. We shed our sweatshirts and laugh at the prospects of a warm, pleasant day.

There is no reason to think it is going to be a big day. Historically, the peak is still more than a week away. After one more pick and a few hundred pounds of fish, we leave our nets and motor upriver to visit Club Ohio, the group of set netters who live across the river from Egegik.

We arrive at their camp an hour before high tide. Gladly, we strip out of our waders and join the gang around the table for a round of beers. A large portion of one wall has been recently repaired; a grizzly bear went on a rampage last winter and tore into every cabin on this side of the river. In some cabins, the bear came through a window. In others, it broke down the door. In the Club Ohio cabin the bear walked through one wall and out another.

At first we all talk about the lightning, but soon Gabe, the talkative one of Club Ohio, begins telling stories of how last year the Woodbine cannery went bankrupt and stiffed them out of $30,000. I ask if they will ever see the money.

"Fuck that. I don't want the money. I want the bitches who own the place. I will gladly give my share to see them beaten to death."

Soon he starts with stories of drug deals gone bad in Anchorage, peppered with the momentary fits of anger at the owners of Woodbine. We all have another beer and I am grateful to be inside and laughing instead of fishing. Still, every few minutes, Carl or I look outside at Club Ohio's nets. There are no fish hitting the nets, and the corks are still showing, an indicator there is no action.

Around six o'clock we get back on the water, laughing and feeling slightly buzzed. This lazy, warm feeling ends the moment we pull up the first net. Expecting a few hits, Carl and Sharon leave me to pull the net over the bow. After attempting to grab the net from the water, I realize I can't lift it. Carl races over to help and we manage to heave the net over the bow and into position, giving us our first look at the number of fish in our net. We can see only the first ten feet of the net, but it is plugged with fish. Knowing we have six hours to go before closing and another net to deal with, we pick the net as fast as we can. Within an hour and a half we fill Sharon's boat with more than 7,000 pounds of fish.

After we drop the net back in the water, Sharon and I jump on Carl's boat, which we anchored up near Sharon's site earlier. Carl takes Sharon's boat to the tender. The idea is to deliver now, allowing us to have her skiff ready for the last pick of the opening. Sharon and I race up to the second net. With all my strength I pull the net over the bow and we begin picking. I try to avoid it, but it's impossible not to look at the shore, reminding me how far we have to go. And each time I look up I see the floats slip farther under water as thousands of pounds of fish pull the net down.

Soon Sharon and I, sweating and panting, realize we can't use this boat to get the plugged net out of the water. The side rails are too high to handle the poundage of fish per yard of net. At one point she and I are butts down in the boat, our hands locked on the net, and pulling, using our legs as leverage. It is no use. So we decide to slow down, save our strength, and wait for Carl to bring back Sharon's boat, which is low to the waterline and easier to work from.

By the time Carl returns, we've stopped talking. The clock is running out. We have to get our nets out of the water by midnight. After switching boats, all three of us begin pulling net and picking fish. It hurts and will for days to come, but we feel in control. By midnight we should have both nets on board, a huge load of fish.

The trouble begins around 10:00 P.M., two hours before the closing.

At first it is just a tug, almost as if the boat has hit a momentary snag. Then the net slides a few inches, burning my palms. Seconds later it heads out of the boat with a fury. Something is wrong.

"Cork the net," yells Sharon, referring to the action of stopping the net from moving. On board every skiff is a single foam cork with a piece of rope knotted to a carabiner clip. To "cork the net" a person has to circle the running net line with this rope and then clip the carabiner to something fastened to the boat. The line will stop once the single piece of cork, which is tied to the carabiner, can no longer move, thus halting the forward motion of the net.

My eyes scan the boat, but I don't see the corking device. The only other way to stop the net from flowing out of the boat is to use all of one's body weight to force one of the corks attached to the net against the metal lip of the skiff, effectively cleating the net by using the friction of the rope atop the cork against it. The tension keeps the net in place, but if the force on the net is strong this won't last long. Trying to grab the net, I turn to see Sharon at the back of the boat pressing her knees and hands against the net and the stiff sidewall. I move toward her.

"*No.* Help Carl cork that end."

I turn and leap over the fish, which fill the bottom of the boat up to my thighs. I reach Carl, who also has all his weight pushing down on a cork, against the railing, which stops the net from flowing out of the boat.

I take a deep breath, enough time to see finally the severity of the problem. The net is coming undone from the shore, and all the anchor stakes are pulling out of the mud. The tide is ebbing and if the net releases from the shore in the next few minutes the free section will drift downstream, pulling us with it. This is a bad situation.

We stop moving for a moment. In our silence the squeaking of the Styrofoam cork becomes a haunting reminder that our gear, our boat, and our lives may be in danger. The cork is being compressed by both the body weight of Carl and the immense forward force to fly out of the boat. Above, seagulls hover, looking for an easy lunch. In the boat the occasional fish flops at our legs, giving out its last gasp of life.

"We've got to pick the rest of the net before it breaks free from shore," yells Sharon, pointing to our right, to the portion of the net anchored in the river and full of fish For a few minutes we hold our position and pick, faster than we've ever picked before. Looking back, this brief moment of stability most resembles what it must feel like to be a passenger in a car half dangling off a cliff. There is a momentary calm before the inevitable disaster.

Then, with a loud snap, the section of net attached to the shore stakes breaks free.

"Get the net off the boat. Now!" yells Sharon.

Carl and I jump over the net, scramble to the front of the boat, and push the net off the bow. At the same time Sharon quickly starts the motor and repositions the boat so we can relocate the net, which is now flowing freely in the river.

"Fuck, find it," she says. "I don't want it in the prop."

After less than a minute Sharon spots a cork and guns the engine.

"Grab it," she yells.

Carl and I grab the cork and begin to pull the net into the boat, inch by inch. Yet, with each foot we bring on board, it becomes clear we are taking on too much weight.

"Gotta move the net to the other side, Carl. We'll swamp it otherwise," says Sharon. I look down and see the edge of the skiff-side barely above the water.

Carl and I reposition ourselves and begin to overhaul the net to the other side of the skiff, which allows the boat to rise a few inches. That done, we immediately return to pulling the net back into the skiff.

"Oh fuck," says Carl with a sense of desperation.

I look up and see what Carl is pointing at. Our second net has also disappeared, and Carl's boat, the one anchored upriver, is now moving. Downriver, toward us.

It can only mean one thing.

"The net has got the anchor line," he says. It won't take long before the anchor gives and the net becomes a lethal snag, racing downriver.

"We got to finish this and get up there," yells Sharon.

And then, the motor dies. This makes the net start to violently pull out the front of the boat. Carl and I jump to the side. Something on my shoe catches for a moment and I am being pulled overboard. Carl corks the net, but I can tell he doesn't have long. The weight is too much. I untangle the net on my shoe and am free.

"Go, go!" I yell. Yelling now becomes the only way of communicating. The heart races and the throat constricts due to adrenaline and panic.

Finally, Carl and I find a way to cork the net, our bodies leaning on a single round piece of Styrofoam.

Sharon digs under the bench seat and pulls out a horn and a life vest. She begins to call on the horn, signaling distress. She waves the life vest above her head, trying to signal our neighbors downriver. They never look up. At this point I am sure we are going to drown.

"Now what?" I ask, breathing hard, my eyes wide with panic.

"Well, I don't fucking know yet," yells Sharon, also breathing hard. She uses the horn again and keeps waving her life jacket.

"Where's David?" asks Sharon, looking upriver toward where her brother fishes.

"I guess they hit low," says Carl with regret. "Never saw it." He's referring to us sitting at Club Ohio a few hours before. The fish hit the net toward the bottom, which didn't show on top of the net, where we could have seen it. The entire time we sat at Club Ohio drinking beers, looking at the nets bobbing on the water, the fish were plugging up the bottoms of the nets. If we had known we would have been working the nets continuously, allowing us a head start on the load.

"Carl, we have got to get this net in the boat. Can't let the other net wrap us," says Sharon.

So, again, Carl and I begin to bring in the net, but with a renewed sense of purpose. Out of the corner of my eye I see another skiff coming our way.

"Hey, can you guys help us?" asks the man. He has come downriver looking for help. "Our nets snapped and we can't get them out."

"We thought you were coming to help us. We've got the same problems," says Sharon, trying to sound calm.

The man leaves, going off to look for help.

"Well, it's nice to know we aren't alone," says Carl as we reach down and haul in more net, full of fish.

Now, because of the tide, which is ripping out, we are parallel to the shore, instead of perpendicular, the normal way to fish a net. We are pulling a net weighing several thousand pounds upriver against a four-knot current. The net is heavy but the current on our boat is even heavier.

All this, and we still have to figure out how to get the other net out of the water. The last ten feet of net take almost fifteen minutes to overhaul.

Carl and I collapse on top of the net.

"You guys are doing great. But we got to get the other net," says Sharon, trying to encourage us.

Meanwhile, Sharon has got the engine working.

With Carl lying on the bow looking for the other net, Sharon drives upriver toward Carl's skiff, which is slowly drifting downstream. We know the second net is wrapped around the skiff's anchor line—that is what's dragging it downstream—but the net could also be behind us, under us, or in front of us. Out here a loose net is a deadly booby trap.

Carl finds the other net, grabs at some line, and after a few minutes determines the net is around the anchor line upstream of the bow. We tie Sharon's skiff to the back of Carl's skiff and climb on board. Carl's boat slips, a result of the added weight of Sharon's boat, but finally it settles and Carl busies himself trying to free the line.

My arms begin to tighten up and my back feels as if I just did five years of hard labor. I know there is another full net to pull in and I try to imagine how I'm going to find the strength to do the task.

The problem of the high side rails on Carl's boat comes back to haunt us. And this time we have no choice but to haul the loaded net up and over them. Furthermore the net is pointing downriver, so we will have to pull upriver as well. Carl takes the corks and I take the leads.

"Where the fuck is David?" yells Sharon again, looking upriver for her brother.

It's almost midnight when Sharon's brother comes downriver, his boat barely above the waterline. He's full of fish. He has Malibu Marty on board. Dave idles just off our bow as Carl and I continue to heave the net into the boat. This goes on for a few minutes. Dave does not say a word; instead he lights a cigarette and watches. I'm about to scream when finally Sharon says something.

"Any help would be really appreciated."

As a response Dave drops Marty on our skiff and then motors off to a safe distance away from us and continues to smoke his cigarette. Dave knows we are in a tight spot, but that doesn't matter. The way he thinks is simple: he's done his work and that's all that counts. As much as I want to focus on the task at hand, I find myself being distracted by the deep anger building for the man smoking in his boat, watching us tear out our tendons.

Anything can happen on the water and the conditions change fast. Worrying about the next problem is not useful. Concentration on the one directly in front of you is the way to stay effective. The amount of fish on both skiffs is now becoming a problem. Both boats ride dangerously low to the water. As Carl, Marty, and I pull in the net, Sharon pitches fish from Carl's skiff to hers, trying to free up room for the net as it comes on board.

Our primal screams of pain and struggle are now loud and out in the open. Pride is long gone, and Carl and I are screaming with every pull. And then there is Dave, finishing off his cigarette no more than twenty feet away.

I have given up on Dave, when Marty suddenly screams, "Dave, I can't do this alone!" He almost breaks into tears saying it.

Carl stops for a moment and looks at Marty, unsure what to make of him.

Sharon looks at Carl and then stands and looks at Dave. "Dave, we could really use your help on this."

Dave Hart's skiff has the words *River Rat* welded into the side. He's four years younger than Sharon and has been fishing in Egegik every summer since early childhood. This village is where Dave has created a niche for himself, by becoming known as the man who works alone. In a place where this can be risky and even foolish, Dave takes pride in this distinction. When asked why he likes to work alone, he will say because nobody can keep up with the speed at which he works. What Dave doesn't say is that no one wants to work with him because he is an asshole most of the time.

Besides being a yeller, Dave is also never wrong. At first this can be hard to accept and makes getting along with him almost impossible. He's annoying and stubborn, and works very hard to be his own hero on a daily basis. Conversations end up being dissertations, no matter the topic: the law, war, politics, fish, the history of the world. And, like a petulant child, Dave always has an answer. *The* answer. But over the years I have learned to accept the rules of engagement with Dave. And the truth is I like him. I'm not sure why, but I suspect it has to do with there being something very honest about his desire never to grow up.

All that aside, not many in the Egegik fishing fleet will dispute that Dave Hart is one of the best fishermen in camp. Pound for pound he can outfish anyone on any given day, no matter the size of their boat or crew. He knows the rules, all the smallest details of the fishing regulations. He knows the river, if not as well as Carl, then very close. More important he understands the river, and knows the tides in his sleep. He's rarely surprised and never underprepared. When I'm in his boat I feel completely at ease. He's a loudmouth, a show-off, and a prick in so many ways, but he is a safe captain.

He is also one of the strongest men I've ever met.

Less than thirty seconds after Dave joins us on Carl's skiff, his feet are firmly planted against the side of the boat. He grabs the leads and corks and pulls both lines together. He screams from his deepest spot and keeps pulling. After he brings in three feet of net he reaches down into the water and grabs another section.

After catching my breath, I join in, also screaming with every pull of the net. Dave pulls more net than any of us. He also coaches us, encourages us, and demands us to succeed. He almost wills the net on board. In fifteen minutes the net is on board. Without doubt it is one of the purest exhibitions of raw strength I have ever witnessed.

• • •

By the time we get our gear out of the water it is 12:30, half an hour after closing. On any other day this would result in a hefty fine, but today the state troopers are preoccupied with the drifters out in the bay, at the northern border of the fishery. In Egegik alone, 950,000 fish were caught on June 26, 2005. It was later relayed to the camp that one helicopter pilot said the river was so full it looked like herring. Nets were plugged up and down the entire river system.

Dozens of nets broke that day. Drift boats were overloaded and almost sank. A few skiffs on the outer banks swamped and were lost. No one lost a life, which with all the nets dragging downriver, was a small miracle.

Twenty hours later, when I wake, I walk to the clinic, my arm so limp I must hold it up with my other hand. The doctor at the clinic says I have severe tendonitis, as he lightly touches the purple bulge the size of a golf ball that has popped up on my forearm. My thumb is completely unusable. The swelling is severe and the pain such that I pace the room and breathe in short gasps, like a child hoping for it to go away. I ice it and take the steroids the doctor gives me, along with strong doses of ibuprofen and codeine.

For the next three days we don't fish. No one has the strength. Instead sleep comes in huge doses of twelve- to fourteen-hour waves, occasionally rising to eat, have tea, or check on the weather, which has turned from bad to worse. The wind is back, blowing 40 miles per hour and stronger.

One afternoon I venture down to the docks, my arm more swollen and purple than before. I ask a Native captain about the lightning and hail on June 26.

"I've never seen anything like that in all my life in Egegik," he says.

Wondering what the ancient myths and shamans would say about the lightning, I ask him if he believes it has any significant meaning.

"Global warming," the man replies, his Native inflection making him sound as if he is chewing on a sack of marbles—slow and thick.

"And the fish? Did you have a big day on the twenty-sixth?" I ask.

He shakes his head from side to side and smiles.

"Too many fish," he says.

In truth we never recover from the day our nets broke loose. It affects how we fish the rest of the season. From then on we become a bit gun-shy, and if the day looks like a big one, we put out one net instead of two, mindful not to put ourselves in that position again.

Late in the season I run into Gabe from Club Ohio. We are both on ATVs, going in opposite directions, both wearing waders and carrying gear, getting ready for an opener. We slow down as we approach each other, and before I can say anything Gabe shakes his head.

"Holy Christ. I don't ever want to have another day like that again in my life. That, my friend, is not worth it."

For perhaps the first time in my commercial fishing life, I know exactly what he is talking about.

And Dave?

So many times over the years I have watched Dave standing at the window in Sharon's kitchen. He just stares out the window, saying nothing. I used to imagine what might be going through his head. Then one time he mumbled something that quelled my curiosity.

"It's the most beautiful sight on earth," he said, talking about the river.

Like an athlete living his high school glory days his entire life, Dave is a man largely defined by a single experience: fishing in Egegik. This is the place he asserts his knowledge, his strong points. Have a person like Dave Hart to dinner in the Lower 48 and he won't say a word. He will be out of place. Egegik is perfect for people like Dave, and there are many of them. The isolated world of Alaskan commercial fishing is built on a paradigm of experience: the more you have, the higher you are on the totem pole.

And his pride, like Sharon's, and like their father Warren's, runs deep. He will not forget that he helped us the day we caught 22,000 pounds. He would never accept money as a show of thanks, and besides, Sharon would never offer it. For Dave it doesn't matter. He has something much more valuable, worth more than gold. We owe him.

CHURCH HILL

In Egegik there is hardly a spot from which a person cannot see 360 degrees to the horizon. Still, everyone will tell you, to get the best view go to Church Hill, just up the dirt road from the liquor store, behind the abandoned Alto family mansion. This is where the First Orthodox Church was built, now long gone, and also the location of the old cemetery.

On the way, I stop at the liquor store owned and run by Don Albright, a white man who came here to fish and make his fortune almost forty years ago. At one time he was married to an Alto daughter, and owned the only bar, the only café, and the only phone in town. Today, there is no bar and no café. Almost every house has a phone. So Don decided to make money the easy way: he sells tobacco and booze. He has the only liquor within fifty miles, sold through a small window attached to his rusted sea cargo container with a handwritten sign that gives the hours: 12:15 P.M. to 1 P.M. and 6 P.M. to 7:15 P.M. daily. Eighteen-packs of low-end beer will run you $24 and a fifth of whiskey is $35.

Black Velvet, Jack Daniel's, and Crown Royal are all on the menu. Budweiser and Miller Lite are there too. Boxed wine goes for $25 for three liters. Chardonnay, Merlot, and Cabernet are the choices.

I ask Don how much he makes each summer off his liquor sales. He doesn't say a word, instead he stares at me a few minutes; his eyes begin twitching and he quickly wipes his mouth.

A cannery worker comes to the window with his weekly check, around $480.

"Can you cash this?" says the man, wearing a hairnet.

"You buying something?" asks Don.

"Pack of cigarettes and six-pack," the man says.

"Sorry, can't do that."

"All right," the man says, disgusted. "Give me an eighteen-pack of beer and a carton. Marlboros."

Don hands him the merchandise and the change. When the man leaves I reappear and again ask how much he makes every summer off the fishermen and locals.

"I'm busy right now," he says, and shuts the window.

Half the town needs Don and half the town wants to throw him out. The drinkers cater to him, laugh at his jokes, and allow him a lifestyle that includes winters in a condo in Hawaii. The other half believes deeply that his product is killing their village and their culture.

Either way it's hard to find anyone in town who actually likes the man.

GOING PAST THE liquor store I stand in front of the dilapidated Alto mansion, named after one of the families that settled this town. When it was built, the house was the talk of the town, with four rooms upstairs and a large family room on the first floor. Today it sits derelict and the tundra grass has grown halfway over the top of the broken-down truck rusting next to the house. I walk on, up the path, and finally I stand on Church Hill, a flat clearing with grass up to my knees. On a sunny day a person can sit on the edge of the bank, with his feet dangling over the seventy-foot drop to the river, and stare across the water to Coffee Point, or straight ahead to Sand Point and out to Bristol Bay, the shallowest part of the Bering Sea. During the long days of summer when the sun doesn't set until after 2:00 A.M., and if there is a slight breeze to fight off the bugs, it can be a good place to stare out at the world. Maybe that is why the original church and cemetery were built here.

There are no markings that identify the old grave sites, just grown-over holes where the bodies used to lie. They were moved after a few harsh winters began to erode the bank and people had begun to notice something eerie: coffins were sticking out of the bluff, a few feet below the topsoil. Human bones dangled from the cliff, sometimes tumbling to the beach below. Most of the exposed graves dated back to the early 1900s, when disease wiped out large numbers of the Native population.

ONE YEAR, before the season starts, I take a walk through the new cemetery, located behind the Baptist church, not far from the post office. I come to a house I know is empty and peer through the windows. There are open cans of food, papers, magazines, and rotting clothes all piled to the bottom of the kitchen window. The trash almost conceals my vision of a small walkway; waist high through the debris to the bedroom, also full of trash. There is a padlock on the door. Later, when I ask about the house, a local tells me that the man who used to live there got another home and decided just to leave his mess and move on.

In the graveyard the oldest graves are nameless but decorated with Russian Orthodox crosses—Christian crosses with a crook in the middle like a sideways sculpture of a bolt of lightning.

The newest graves are toward the front and have colorful decorations surrounding them. Baseball caps, fishing lures, and net corks are among the favorite knickknacks for the dead. I notice one gravestone in particular. It reads: Evan Everett. Born August 21, 1962. Died September 14, 2004. This is the grave of a man known his entire life as Boobuck, a thin, quiet, and kind local who lived near Sharon. He used to come over late at night, usually drunk, with his partner of eighteen years, Darlene. They would ramble the Native ramble. One of them would start with how many fish there are and then casually switch the topic to what some neighbor had done to them. Most of the references were so local that it was difficult to know what they were talking about, and thus one took clues from their body language about how and when to interact.

Boobuck died in a way that is typical for this town: tragically and suddenly. He had been drinking, enjoying the dog days of summer and smoking fish with his friend Ben. "I told him we didn't need any more fish," says Ben when I ask him about Boobuck's death.

Boobuck thought otherwise and went down to the river in his waders to check his subsistence net. Boobuck walked one inch too far and once his toes were off the bottom of the river the waders became buoyant and pushed his feet up and his chest down, allowing the water to flood in. Too drunk to fight the current, Boobuck disappeared in seconds. His friends, who reached out for him, say he was gone "just like that." They found Boobuck the next day on a sandbar.

"I don't know," says a Native man when I ask about how many people have drowned here over the years. "But no one in Egegik knows how to swim." Of course this makes sense, because swimming is impossible in this river.

From birth Natives learn the river. They know the tides and have great respect for the powerful mood swings of the river. Yet they rarely speak of it and they don't trivialize it. The river mostly remains a silent topic, a quiet energy that pulls at their souls.

One man, who was born in Egegik and has been fishing these waters for almost sixty years, tells me he wakes up every night of the year at the top of the tide. During the fishing season he walks down to the river and checks his anchor and the flow of the water. I asked him what he does in the winter, when he's not fishing but the tides keep turning.

"Just wake up, turn over, and go back to sleep."

The people of Egegik do not intellectualize the river in terms of how it affects their lives. They won't talk about how the river and its massive tidal flows loom over this village like a giant sunspot, pulsing through their collective psyche. Instead they speak of it as a resource, a thing they deal with from time to time. Yet I'm quite sure the river is to these people what the moon is to wolves, a physiological magnet pulling at the salt that flows through their blood vessels, the liquid in their brains, and the water in their plasma, to and fro.

After all, as fetuses we live nine months in water, gilled if you will, in our mothers' wombs. Then, born liquid sponges into a dry world, choking for air, we spend the rest of our lives rehydrating. So imagine a place where every twelve hours the tide changes the waterline by as much as twenty feet. Where the flood of the high tide wakes up one man every night for almost sixty years, and where the rip of the ebbs tugs at the deepest places of the psyche.

Walking past the cemetery is a group of cannery workers who, like me, are waiting for the fish to arrive so they can work. They are black, not American black but the deep black that comes straight from Africa.

All men, each wears a green head net to protect himself from the swarming mosquitoes. Talking quietly in a foreign language and dressed in button-down short-sleeved shirts and pleated pants, they stop for a moment to acknowledge the cemetery and then continue walking, talking.

I ask one of the men where they come from.

"Sudan," he says.

Since the beginning of commercial fishing in Alaska the fish companies have imported foreign workers. First it was the Scandinavians and Italians as fishermen, along with the Chinese and Filipinos as general labor and cannery workers. Filipinos still remain a large workforce, as do Vietnamese. Lately Mexicans have made their way north for the season, as well as the standard workforce of young college kids from the Lower 48. For these Sudanese it is likely that the cannery has arranged with the government of Sudan an exchange of cheap labor for work visas. Last year it was the Turks. All I know about the Turks is that they were never paid, and it took the Turkish embassy's pressing charges finally to get the cannery to pay for their tickets home.

Thinking of this story I ask the man from Sudan, who, like his countrymen, is tall and lean, whether they have ever eaten salmon before. They haven't.

"Where are you from in Sudan?" I ask the man as we stand near Boobuck's grave.

"Darfur," he says with a beaming smile. "Have you heard of Darfur?" he inquires with great anticipation, perhaps responding to my nodding head.

I tell him I have and that's all I say. I want to tell him so much more, like how I'm sorry the world watches while his people are exterminated. I wonder if this man thinks the reason no country is willing to help is because they are black. African and black. African and black with no oil to fight over, no food to grow, and a country the size of Alaska filled with sand. I wonder if he is glad to be here, in a windblown village on the edge of the world.

"I don't like the bugs," he says.

He rejoins the others and they walk on, strolling really, as if enjoying a Sunday walk together, something I suddenly realize I have never seen in Egegik. The locals here don't walk anywhere they can drive in their four-wheel ATV Hondas, and are rarely seen in groups. I can't help but think that no matter how far Sudan slips off the world's political maps, Egegik is even farther off the map, culturally estranged from the world.

I look at a few more graves. Norman "Pete" Olsen Sr., born April 24, 1972, and died January 6, 2003. The tombstone reads, "Beloved son, brother, husband, father, friend." Perched on the cross over his grave is a hat imprinted with the words "Better red than dead, 2003." And dangling off the edge of the cross is a halibut hook and a Velcro strap, a piece of fishing gear of some kind. This was Carl's best friend, the one who took him on his first caribou hunt.

Pete died of what could best be called consumption, not the kind found in Dickens's novels but a more modern take on the disease—alcohol consumption. He died of living large, drinking and eating too much, and bloated organs. From all accounts he was a happy person whose priority in life was to take care of his wife and kids, not easy in a place like Egegik. Pete left behind a young wife, Corrine, and four children. He also left behind a boat, the *Gunsmoke*, which Corrine wants to get in the water for the season, to help pay off her debt.

So why does Corrine remain in Egegik, where life without her husband will only get tougher? The only sensible answer is the same reason everyone else stays here as well: they stay because this is where they are from. They stay because many don't have the means to leave. They also stay because to leave means dealing with the outside world and they are not sure they want that. Presumably, no matter the details of why they have to stay—poverty, family, children, the next big payoff—they end up staying because it is what they know.

Also, for many the outside is corrupt and evil. Many Alaskans feel this about the Lower 48. For them, anywhere outside Alaska is a wasteland. The Lower 48 is a place where the government, the police, the local school board, and everyone else is out to screw you. When I mention to people here that Alaska is no different and has a history of being run by greedy politicians in bed with big oil and timber corporations, they laugh and acknowledge it as true. "Yeah, but we're free here."

Free. Not a word I associate with Egegik, at least upon first impression. But over the years I think I've come to understand what *free* means to a resident of Egegik. In one sense they are free from dealing with the world most people take for granted. They are free to live in a social setting that resembles a large family, where the laws are stretched and social norms distorted. Life is simple. They are expected to survive the winter and fish in the summer. They are free to live in a place on the edge of the wilderness, where no one has to imagine what it means to "get back to nature," perhaps the most freeing sensation of all.

I continue to walk the graveyard, pushing back the tall grass that conceals most of the names. Some plots are so small I guess they must be children. This is the quiet tragedy of Egegik. I'm not sure of its source, only that it can be felt on any given day, when the wind is still and I look around for the young people. There aren't many. The kids are either shipped out for schooling or they stay, to live in this isolation and, it is hoped, to avoid dying young.

I push back some more grass and see the name of Pancho, a young friend of Sharon's, whom she met in the early days of her time in Egegik. He was killed when he lifted a pipe to look for a squirrel. The end of the pipe touched an overhanging electrical wire and electrocuted him.

There are even more ways to die in the Alaskan bush. Fermented salmon is an old Native delicacy, and like the poisonous blowfish in Japanese culture, it entails a certain amount of risk to get the flavor just right. First, they bury the salmon for an extended period. Then they recover it, hopefully in time. But sometimes the fish has developed deadly botulism. Every year I read in the Anchorage paper of Native people becoming sick, and sometimes dying, from eating rotten salmon.

In August 2005, this report prompted a medical alert: "The state Division of Public Health has reported 2 recent botulism outbreaks in Interior Alaska villages, but it hasn't named the villages. Health officials sent nurse practitioners to 2 separate Yukon-Kuskokwim Delta villages Mon, 22 Aug 2005. They went to help with 4 people who became ill after eating fermented salmon, a traditional village meal."

For some historical perspective, I visit Paul and Nattie Boskoffsky, whose house sits on the bank of the river, with perfect views of Goose Point. The dirt road from their house leads straight to the graveyard, where their daughter rests in peace. Outside their home a few nets lie stretched on sawhorses, ready to be mended. Otherwise theirs is a well-kept yard. Inside, the house resembles a typical suburban home; clean, with a rug, couches, and a kitchen with a dining table. It's roomy with lots of window light.

On the living room wall hangs an oversized black-and-white photograph of Paul's grandfather Nicolai Ruff, the chief of the Kanatak, an ancient village on the Pacific side of the peninsula, where Paul was raised.

I ask how Egegik has changed over the years.

"Alcohol is the problem here," says Paul, economical with his words.

And the outsiders who come here every June to fish?

"I don't think they do much for the town. In fact they take from it. Yes, we get a tax, but that's all. They bring the alcohol and drugs. And they are aggressive fishermen. The gentlemen's fishing is long gone."

Nattie smiles, as if used to Paul's complaint about the outside fishermen.

"I think they bring a work ethic to our town that does not exist in our young people today," she says. Paul's expression does not alter. Nattie seems amused, light at heart. "Kids here don't see their choices. My dad was a drunk but I chose not to be. I always say, 'You gotta change the water hole.' "

A young crewman knocks on the door. The tide has changed in the last forty minutes and a line has wrapped itself around Paul's skiff. Paul stands and shakes my hand, thanking me for stopping by for a visit. He asks me to say hello to Sharon.

Nattie keeps smiling, talking to Paul about logistics while I take a moment to look at the pictures of their daughter, Jewell, scattered throughout the room. I met their daughter the first year I fished in Egegik. She was beautiful, young, and shy to outsiders. She worked as crew on Paul's boat and was well liked by the entire town and camp. Then one year she got sick and soon after died of leukemia.

Leaving their house, I walk toward the post office. It's raining and the wind is blowing, as usual. I think about Jewell and her death. It seems almost normal when compared to other deaths in Egegik. She got sick and died. She didn't freeze, get electrocuted, drown, or blow up her liver with booze. She just died too young.

THE EVENING IS still and as the wind drops the temperature rises, bringing out the bugs. It is the Fourth of July and families, fishermen, and strangers gather on Church Hill to light fireworks. A few hours earlier a man with a distinctive Russian accent stopped me on the road. A fisherman, he was obviously lost as he wandered Egegik for the first time.

"Where is the liquor store?" he asked, his face parched, looking as if he'd walked fifty miles across a desert. I pointed behind him, toward the cannery, and told him it wouldn't be open for another hour.

"Any women?" he continued.

"Sorry, on your own there," I said, and kept on walking.

This marks my third season in Egegik, three summers spent in a place I find difficult to describe to others, and that took me a year to pronounce correctly. Carl and his wife and kids stand nearby laughing and staring out to sea. There is whiskey and beer and talk of tomorrow's opening. On the river dozens of drift boats head toward the mouth of the river, riding out on the last of the tide, making sure they don't get stuck in the mud during the ebb. I stand on Church Hill, a beer in hand and one more in my pocket. I squeeze my sore fingers, feeling a newfound strength, perhaps from the alcohol. I tell myself I can do this.

I can live here. I can do this life. It is almost a dare to myself to stay a winter and disappear into the Alaskan life of tundra, wind, and snow. And why not? Why not disappear into a life dictated by the ancient rhythms, where the salmon and bears and ice floes dominate the daily talk? I can listen and learn from the locals how to track animals and become a better fisherman. The river will become my local road, something I know how to navigate in the darkest of nights. I will spend weeks up at the lake where I will hunt caribou for food and occasionally kill a moose and share it with others. In time I will disappear from my old social networks and no one will inquire anymore where I am, and if they do, friends will say he lives in the wilderness. The wilderness: where everything is still possible.

Carl's son Griffin lights off a firecracker and everyone claps. Feeling bold I lean over to Carl, who is all smiles for his son, who I notice looks more like Carl than his Native mother.

"You going to raise Griffin here?" I ask.

"I don't think so."

"Why not?"

"Too hard a life."

I step back and stumble, falling on my backside. I have stepped into one of the old graves. I can imagine all the stories of these people's deaths. How many died of exposure, passed out from drink in the snowbanks? How many went overboard and were never seen again, their graves filled with an empty coffin out of respect?

"What does 'Boobuck' mean?" I ask Carl, remembering the gentle man who used to visit Sharon.

"Shitty Pants," he replies with a grin, offering no explanation as to why he had this nickname.

Just then I think of the men from Sudan and all their stories, and wonder what they will tell of this place when they go home.

So THE GRAVES have been washed away, the bones sucked out to sea where they settle into the channels and sandbars: battlegrounds for the salmon, all churning over one another, biting and killing each other. High on the bank we can't hear their struggles. Up here it's quiet. We eat and laugh over drinks. And later we will sleep, but they are down there, the bones and the fish, mixing together, spawning tomorrow's generation.

I think of the tides, how they pull at us. Most of the year, I live in the desert, a dry sea of sorts, with its own cycles of oversized moons and monsoon rains, tugging at my desire to wander the land of mesquites and disappear from time to time. Still, every year, I, like so many others, return to this place where the daily tides rule every aspect of life. Maybe coming here *is* like returning to the watery womb. Or maybe we are like the salmon, called to run upriver, against all odds, just to spawn, and then die.

Out on the river a sockeye salmon jumps, his chrome skin shimmering against the long, golden rays of the Alaskan sun. It jumps again, this time reaching higher, its body wriggling and bucking before slamming back into the water.

SEEING NATIVE

STANDING IN SHARON'S ROOM, I AM LOOKING OUT THE WINDOW at the river. Next to me is Scott Olsen, a Native who for many years had lived down the road from Sharon in a cabin with no electricity, no plumbing, and no water. Recently, after having a baby, he and his wife, Michelle, the same Michelle who worked on Sharon's boat my first day of fishing, moved from their shack into his mother's house.

Scott's scoping out the drift boats with Sharon's binoculars as the captains begin their sets, drifting upstream with the flooding tide. Like a commentator at a sporting event, Scott announces the play-by-play moves, his voice rising and falling with great anticipation of each maneuver.

"Ron's ready to put his set on the bar. Waiting. Waiting. Bang, there's the buoy," he exclaims, with more than a twinge of envy. He quickly swivels the binoculars to another boat. "There's the set, over there," he yells. "That's the one you want. Why is he still using his engine? Save gas and drift out the set."

His speech is quick and confusing as he jumps from boat to boat, giving rapid-fire advice on how he would do things, if given a chance.

PUT YOURSELF IN Scott's shoes. Scott doesn't have any money, not even enough to buy a can of food. His freezer is out of moose and the caribou didn't come this winter. He keeps telling anyone who will listen that he needs to start freezing fish for the winter, but it's halfway through the season and he still hasn't put a boat in the water. Ask why and the answer shifts from one thing to another: his engine doesn't work, his boat is on land, and he has no crew.

Later the same evening he sits at Sharon's kitchen table devouring a meal of chicken and rice I cooked earlier, making sure I made enough for the usual parade of people that comes through Sharon's house on any given night. It's not a burden for Sharon; this is her extended family.

I sit near the stove, whiskey in hand, and try not to watch, but I hear him tearing and gnashing at the bones, slurping up the rice grains, piece by piece. I've seen this kind of ravenous consumption: in war, and a few isolated places around the world where supplies are limited. Scott's not a slob. He's starving.

When you first meet Scott it's hard not to notice his hollowed-out cheeks and emaciated body, clear indications that he doesn't eat enough. And his long mullet peppered with gray makes him look much older than he is, thirty-six.

Occasionally he pauses while eating, just long enough to moan, "Hmmmm, this is so good."

Then he puts his hands together in prayer and mumbles to himself with his eyes closed.

"Haven't eaten in a while," he says, laughing nervously.

Scott's considered Native, although whatever white blood flows through him wiped out many recognizable Native features. Instead of black hair and dark eyes, his hair is dark brown and wavy, with a touch of gray. His eyes are deep blue and buzzing from side to side. He walks fast and talks even faster. And when he speaks he laughs, and then proceeds to rebut almost everything he says, as a precaution, just in case it all sounded strange to the listener. He chain-smokes cigarettes with deep conviction, like a prisoner on death row cherishing his final draw. And he drinks lots of soft drinks, which may account for his set of rotting teeth.

Scott has a dream to be a captain of his own drift boat. The problem is that he doesn't have a drift boat, only a skiff. Also, he doesn't have a permit or any money. So, he waits. And every season he looks through the binoculars at boats that he could helm, watches other captains make sets he could make better and dreams of the money he's not making.

There's a "Scott" in every town. A man trapped in a dream of what his life could be if only he could get that one extra dollar, hour, phone call, or the right woman. He's a stray dog of a man who can be seen walking the road, and when you offer him a ride he gladly accepts and in return, as payment, you get an earful of his theories on why you and he aren't rich, aren't successful, or aren't lucky. He weaves intricate theories on how the government has a vicious lockdown on everything that could possibly benefit him, us. Of course he has plans of getting rich. These plans range from bagging a hundred thousand pounds, to finding an endangered fish in one of the village ponds, which would then spur the government to spend millions of dollars cleaning up the pond and protecting it, of which he would collect his fair share.

He has the animals figured out as well. The caribou are all gone because there are too many wolves. When I ask about the two citations he received one winter—for killing an underage moose, out of season and without tags—he nods his head and smiles. "*They* don't want the Natives to hunt," he says, sure that I understand the conspiracy implied.

Don't get me wrong. I like Scott. A wingnut to be sure, but at least he's not staring at the television waiting for the laugh track to prompt his emotions. Instead he's dreaming up the next scheme, the next big idea. Who can't like a guy who doesn't get down when none of those dreams pan out? Also, I like that he pays attention to detail. He's smart and extremely well informed about all things that happen in and around Egegik. He can tell you the names of all the birds that migrate to this part of the world and knows the names of most of the boats at the dock and who crews them. He knows the details of the most recent town hall meeting, including which violations have been broken. As for his own hunting violations? The truth is that no true hunter would ever make those mistakes. But Scott is a Native, and for him, in his mind, there are no rules on the land. To Scott the outside fishermen come to Egegik by choice. They choose this job. But for locals there is not a choice. They have to do it or else they and their families don't eat.

"There's more food," I say, my eyes fixed on the pieces of rice dangling from his jacket collar.

"No. I better leave some for someone else. You never know," he says, and then laughs loudly for a few seconds. He gets up and heads for the door, to smoke a cigarette.

"You want to look at some ivory?" he asks me, with a sense of urgency.

Maybe tomorrow, I say.

He stands by the door for a few minutes, talking about his engine, his boat, and the permit he is waiting for in the mail.

Finally I get up and walk to the door.

"Going to camp?" he asks, enthusiastically.

"No. Just getting some air," I say.

"Oh. *But*, if you are headed up to camp I'll grab a ride," he says, his eyebrows moving up and down.

On the ride into camp I am struck by the same thought I have almost every day I am in Egegik: how do people survive in this place, so isolated from the rest of the world? And how can someone live in a house for years with no heat and running water when it's 20 below for two straight weeks? At times, people like Scott seem both out of touch with the modern world and yet fully capable of wanting all that it offers. He wants heat. He wants a boat. He wants to make money, but he is trapped by his own sense of entitlement, almost a sense of nobility. He has a stellar reputation as a crew member and hard worker, but this season he refuses to clock hours for anyone else. He has decided this is the season he will become a captain. Or starve waiting.

He won't fish for locals as crew because that would mean he hasn't succeeded on his own. He won't fish for outsiders because it would be an admission to him and locals that he can't get his captain's card. So, his self-inflicted ideas of being a captain have trapped him from making a living. This is not unusual in a place like Egegik. For Scott and many other young men, being a captain of your own boat is the pinnacle of village life. It reflects a sense of accomplishment and announces to others that you have arrived.

This conflict in Scott illuminates a much larger issue that haunts the men and women of Egegik. It explains why outsiders feel slightly incongruous in Egegik, as if standing in a time and place from long ago that has suddenly been dropped into modern society, resulting in something that doesn't quite fit. This confluence of two worlds is not a mistake. Scott's seemingly conflicting ambitions are not a mistake. They are a direct result of an ancient people mixing with a modern world, a phenomenon that creates a cultural rift between old and new.

It all began in Egegik when the Russian fur trappers began marrying the local women.

Unlike many Native American tribes in the Lower 48, who still want nothing to do with white culture, most Alaskan Natives have adopted various aspects of the outside culture without much resistance. Some may say they are selling out, or unjustly put upon. Maybe. But there is another, more likely, explanation: the extreme land on which they live does not allow people to resist overwhelming outside influence.

When white men came west across the American continent they encountered indigenous people living in tribes of loosely organized encampments or villages bonded together by blood and geography. Faced with an overwhelming force, and determined not to capitulate, the Native resistance led to their eventual massacre by U.S. soldiers. That they were organized by tribes, complete with alliances with other tribes, only led to their death by the thousands.

In Alaska the strategy of conquering the Native peoples was different. The Russians spoke of conquering the indigenous people not by tribes but one village at a time. Why? In the more temperate zones of the Lower 48, Indian tribes had a luxury that did not exist in Alaska. They had weather that did not kill them the moment they stepped outside. The southwest pueblo tribes had secret hiding spots in box canyons they could rely on for months on end. They could walk one hundred miles, following animal trails through the rugged desert mountains, eluding capture. They practiced such tactics for decades, staving off the overwhelming numbers of the U.S. Army. If trouble came their way they signaled to one another, sending their best warrior on foot to warn others.

But in Alaska there is nowhere to go. For nine months of the year, the elements kept people from traveling outside their villages. And in the summer people were too busy fishing to travel, getting their supplies ready for the oncoming winter. They could not hide in canyons, or ambush with deft warrior ways. They could not communicate with other villages and form unions of resistance. In the end the extremes of the land, which had once isolated them from their enemies, worked against them.

The Russians used the isolation of one village from another to their advantage. At the first sign of resistance, the Russians split up the families, usually enslaving some of them in another region. To survive, the Natives assimilated. Over time Russians began to marry local women, or at the least they carried on affairs. This explains why Russian family names can be found in most bush villages throughout the peninsula. The history is not spoken about much by the locals, but genetic hints of their past are permanently sketched into their Slavic jawlines, wavy hair, and blue eyes, revealing a Russian heritage. Later, some Native families would take on a different genetic coding when the Scandinavian men came to Bristol Bay in the early 1900s as pioneers in the salmon business.

Now we are in Scott's net locker, located in the front yard of his mother's house. "The drinking is killing half the town while the other half is trying to keep the place alive," says Scott, who does not drink. He's seen what it has done to friends and family. Pete, Carl's best friend, who died young, was Scott's brother.

"But I guess if the drinkers keep dying that will calm everything down," he says, while furiously searching for something near his workbench, which is covered with rocks, corks, small pieces of wood, and glass jars of ivory.

His mother's house is on HUD Row, referring to the Housing and Urban Development fund that built a string of homes in Egegik. The yard, like so many in town, lies in ruin with car parts, sawhorses, nets, rotting cork lines, kids' toys, a baby crib, moose antlers, caribou antlers, an ATV trailer, furniture, and, of course, broken-down outboard motors.

Scott reaches under his workbench and like a magician working his crowd slowly unveils a piece of ivory. "Dinosaur. Mammoth, to be exact," he says, fondling the piece. "How much money you got?"

"I don't have any."

"Ten thousand years old, that one," he says, giggling and ignoring me.

Reaching for another piece he quiets down, almost to a whisper, his eyes growing large. "You can't tell anyone I am showing you this piece. It doesn't belong to me but I can sell it and split it fifty-fifty with the owner. The problem is that I can't give him the money, he'll just waste it on booze."

"Charlie?" I ask.

He flashes me a suspicious look. I tell him I know Charlie has a severe drinking problem, the kind that would make him sell his last can of food for a drink. I also know Charlie collects ivory.

"Yeah, but you can't say anything." He shows me the piece. It's an ivory cribbage board. Small but intricate, made by a Native, maybe an Eskimo craftsman up north.

"How much money do you have?" he asks, again.

"Scott, I just told you I don't have any money."

"Oh yeah," he says and then puts the cribbage board away. Immediately he turns to grab something from behind him, buried under some webbing. He pulls out an old fur pelt pinned on a board, ready for sale. Then he grabs at some jars and empties them, displaying dozens of agates and rocks he has collected over the years.

"Why do you think alcohol is such a problem in places like Egegik?" I ask, knowing the answer is both simple and complex at the same time.

"Family. Family is all that matters. Without it there is nothing. I think when the family breaks down, the drinking gets worse."

IN HIS MOTHER'S living room Scott opens the mail. The television occupies the center of the room; boxes of old clothes spew out of the two bedrooms near the front door. People who live on the edge of civilization tend to collect things, even items that are worn out or have lost their use. Who can afford to throw a button away in a place where buttons are not available? One never knows when he will need a spare part of an old engine or clothes for a neighbor or a button or a belt or an old pair of shoes.

Outside it's raining and inside it's cold and dark. Gray smoke hangs over the room as he rolls another cigarette and talks fast and furious about whether to fish Egegik or Ugashik, a river fifty miles to the southwest. In the envelope on the table is exactly what he has been waiting for all his adult life, a blue captain's card, no bigger than a credit card. This is the solution to Scott's problem. It is his mother's permit, but she has given him the right to use it for this season only. Of course there is a $30,000 lien on it from banks and the IRS, but in Scott's mind all that matters is that he can, for the first time, legitimately call himself a captain.

"I don't need a crew. I'll go out for two days at a time. Just take a bag of rice and a stove. I don't need an engine. Just a skiff and an anchor. I can drift the river with the tides, and let the boat turn on the anchor."

His excitement is growing by the minute.

"You know I've been all over the world, worked everywhere," he says, fiddling with his fingers rolling the cigarette.

"Where have you been?" I ask.

"Seattle. Washington. Anchorage. Everywhere," he explains with wide eyes and a nervous stammer.

"I'll fish the silvers in August. Make my thirty thousand dollars and be even for next year."

Silvers—also known as coho—are a smaller salmon run than sockeye, and pay half of what the sockeye do. At twenty-five cents a pound, he will have to catch more than 100,000 pounds of silvers to pay off the liens. That isn't going to happen, even with two boats and a crew of six.

"Going to AGS?" he asks as I stand to leave.

"Yeah, sure. You need a ride?"

Back outside, Scott lunges into the weeds and pulls out a brand-new net and loads it into the trailer behind the ATV.

Then he loads another net.

"They owe me a hundred twenty-five dollars for each one. Now I just got to find the guys," he says, referring to the fishermen who contracted with him to make the nets.

As Scott and I head toward town he yells over the engine, "When you get some money, let me know and I'll sell you some of that ivory."

Scott's people have lived here for more than six thousand years. These are the people who tracked the bear and caribou for millennia, long before the rifle was born. They built skin boats to hunt seals and whales. When the pyramids in Egypt were being built Scott's ancestors were building sophisticated wood traps to catch the summer salmon run. As Carl says, no matter how screwed up a local may seem, he can outhunt and outtrack any white man who comes here. And he can do it in the snow wearing sweatpants and gym shoes.

For all the events that have taken place in Egegik, there are no mythical heroes in this culture; there is no written history. The Native language is dying and the young people want to move away. Yet, when not fishing, or exhausted, my mind is constantly in the past. I can't help but wonder how many people died learning that monkshood was poisonous. How long did it take to learn how to track a bear and then kill it with a spear or some other sharpened stick? I imagine there were heroes here, known only in the generation in which they lived. Maybe they were great hunters, or perhaps fearless warriors against their enemy village. I look at Scott and see a man from the past. A person struggling to meet the demands of today's necessities, but with a mind longing for an era when his value would be weighed by how much meat he secured for the winter. Instead he is overpowered by the white man's magic: blue captain cards and fish tickets displaying how much money he makes on a day's catch.

After I drop Scott off, I ride back along the road to Sharon's. The past is out there; I can feel its grip in the bitter cold and see its shape in the tracks along the shoreline. Yet I can't always hear it. I can't hear the stories that are buried in this ground, the stories themselves swallowed by the past and the vastness of the landscape.

Each year upon my return to the Lower 48 people ask me if the people of Egegik feel put upon by outsiders. Do they feel taken advantage of? Do they feel white culture brought them alcohol and disease and took their resources? I can only guess the answer varies from person to person, depending on how well they are tapped into the modern economy. But honestly I don't think they think of it much, and if they do it's only for the few weeks when fishermen overwhelm their villages. The problem is the question itself. To have a proper answer one must assume that the people of Egegik and other villages have the time, energy, or the mind-set to analyze their predicament, specifically in relation to outsiders. That is a luxury few people here can afford. They are much too busy trying to fill their freezers with food, take care of their elders, and hunker down for the coming winter darkness.

My mind races with these thoughts and then it all disappears, in a single flash of exhaustion. My body takes over. The pain and the sleep deprivation are far more pressing than wondering if Scott will get his boat working, or if the Natives resent my presence here.

I shut the door to Sharon's cabin. I wrap my aching hands in an ice pack, open a cold beer, and spend a few moments guessing which local will show up at Sharon's for a late-night visit.

CRIME

ALASKA HAS THE HIGHEST PER CAPITA INCIDENCE OF ALCOHOLISM, rape, and suicide in the United States. In Egegik, incidents of rape are three times the national average. And it is only recently that the men of Egegik go to jail for fondling their daughters and nieces, a dark side of bush life rarely spoken about to outsiders.

"He did something wrong," a man will say, speaking of another man everyone believes molested some younger girls in his family but who still lives in the village. The police have never been a reliable source of justice in Egegik, and besides, there has always been a reluctance to bring outsiders into the village to deal with what many see as a family problem.

The nearest state trooper to Egegik is fifty miles away, by plane. Response time is measured in hours, sometimes days, but certainly not minutes. There has been an ongoing vacancy for a public safety officer in Egegik for years, but there are rarely takers. The pay is between $17.05 and $18.60 an hour and the officer is not allowed to carry a firearm.

HISTORICALLY, CRIMES are often not reported, instead allowed to fester in families over years. Why? One reason is because to air a grievance one must have a forum in which to speak: a court, a police station, a lawyer's office, even a mayor's office. Egegik has always lacked these institutions. The village is like a family of eighty people who may not like each other, but who have to live with one another anyway. What happens here stays here and dirty laundry is not for the outside to see.

This probably explains why the killer of Carol Abalama's mother was never apprehended. She was found in the freezing snow under a building near the docks, in her nightgown. There was no police investigation, even though the only path to her body was blocked by a very heavy barrel of oil, which an older woman, barefoot and in her nightgown, would not be capable of moving herself, especially on a freezing night.

That isn't to say justice is not occasionally doled out, family style. One season I returned to learn that a man was almost beaten to death for molesting a young girl. The state troopers arrived in time from King Salmon to save him from dying.

I know a man who has family in Egegik, but he lives near King Salmon. He is no longer welcome in Egegik, and will probably not visit again the rest of his life. Something happened once—no one will tell me what it is he did—and the family kicked him out.

What is the future of the individuals who commit the unspeakable crimes, such as child molestation? If they stay in the village, these men grow old, all alone, reading two-week-old newspapers in the village council office, drinking coffee and waiting for seasons to come and go. They are walking shells of men anxiously eyeing the cemetery, perhaps hoping that when the end comes they at least get a good view.

And what about the victims of these crimes? Some girls grow up to raise their own families in Egegik, in plain view of the uncles, fathers, and cousins who violated them. Others have relatives in other parts of Alaska and are sent away by the immediate family. And many of the girls in the younger generation are leaving Egegik, before they find themselves trapped. They marry outside the village, some to white fishermen, and leave Egegik forever. Meanwhile the young men are left to stay in a village where marriage becomes more and more unlikely.

THE FISHERMEN TOO have a set of unwritten laws, and like the locals they often don't wait for the troopers to intervene.

The year I worked in South Naknek I learned something about how fishermen deal with crime. One night a cannery worker slipped onto a drift boat and stole money and equipment. The fishermen didn't call the state troopers, at least not immediately. Instead they found the young college student and put a shotgun in his face, threatening to kill him and throw him in the river. Considering where the kid was, there was no reason not to believe them. They got back all their property and money. Eventually the troopers flew over from King Salmon to fill out some paperwork and take the kid to the airport.

In the summers I've spent in Egegik most of the violence between fishermen has been limited to fistfights and exchanges of harsh words. One man, Al, was kicked out of camp for a year for ramming half a dozen boats and running over half a dozen nets. He is a short, muscular man with an angry face, who seems to take great joy in belittling his crew members. The irony is, if his crew doesn't quit, he releases them early every season, trying to save a few bucks, and then when more fish come upriver he skulks around the camp asking people to crew for him. He visits Sharon every season for this reason, and every season Sharon holds her ground and tersely tells him she doesn't know anyone he can hire, and then closes the door.

"He's a dick, and I'm not sure I'd pull him out of the river if his boat sank," she says to me once.

On the water fishermen develop reputations for corking, the act of laying a net down directly in front of another boat's nets, thereby cutting off the fish. At night, as we cross the river, we sometimes hear ongoing arguments in the dark, usually two captains lashing out about who is responsible for tying up their nets.

Many captains carry guns, which they say are for seals. But they also like to keep them handy to fend off other fishermen. Dave Hart tells me a story of an encounter he had years ago with a drifter, a gun, and a machete.

"This local guy wraps his net around my buoy while I am picking my net. I have a crew member on board. I tell the guy on the boat I will pick my net and then help him get his net off. But I tell him I'm keeping the fish in his net [if someone wraps their net around another person's gear then legally they have to forfeit their catch to Fish and Game and replace any gear lost in the incident]. He says no, he wants his net and fish. I say fuck off. So he rams my skiff with his drift boat. I get off the net and race back to shore, where I had a cabin. Got a machete and a shotgun. Went out and cut the fucking net with a machete and he started at me again. I fired two rounds above his boat. He took off and called the cops. I called the cops too. The guy lost his license for three years."

OF ALL THE crimes of fishing and the dark crimes of the village, including molestation and spousal abuse, one crime in particular lingers over Egegik with an air of mystery. It happened August 22, 1984, when a man named Ken Hunter motored out to the confluence of the Egegik and King Salmon rivers and shot two men, killing both. One, Richard Mason, was his former fishing partner. Presumably after shooting them, he sank them and the gun in the river. A few days later, while state troopers looked for the missing men, Hunter left for a vacation in Hawaii. Four years later he was indicted for murder in Alaska.

Sharon's container van, the one attached to her house, used to be Ken Hunter's. She became the owner after Hunter's cache of ammunition suspiciously caught fire one night, exploding it from the inside out, permanently warping the steel.

"He claimed Richard shot at him the night before," says Sharon when I press her on the details of the crime. "I didn't see that, but I was the last one out on the river that day to see them alive."

As the Alaskan authorities lined up their case, Hunter fled back to Montana, where he had once been a psychology professor. Lacking proof to arrest him on murder they arrested him on charges of child molestation of his own seven-year-old stepdaughter, a crime for which he was extradited to Alaska and eventually convicted. It would take almost another year for the state to build their case for murder. The jury deliberated less than two hours and convicted him for double homicide.

"That guy was crazy," remembers Dave Hart. "We were teenagers so it was kind of exciting being around someone so out there, but really he was freaky."

Known as an intense person who liked to spend his time chasing down poachers and professing the virtues of a pure race—his own—Hunter was also a person who enjoyed pushing others to the brink of conflict.

One day I call the director of prisons in Alaska to inquire about Ken Hunter's case. The person in charge tells me he isn't serving his time in Alaska but has been farmed out to Arizona. As we talk about Hunter's case, with which he is familiar, he encourages me to look into some other famous murder cases in Alaska.

"One man, known as the Willow Creek Killer, has been in prison here for almost thirty years. Mass murderer. He would hire prostitutes and then take them in his truck out to the woods, places that he knew very well. Then he would release the women, telling them to run for their lives. Then he would hunt them down with a bow [and arrow] and kill them."

I ask what the man is like in prison.

"He's the barber. The thinking is that his violence was only triggered if the victim was being hunted. In prison he's fine with a razor. He's a model prisoner."

I attempt contact with Ken Hunter a few times, but I get no reply. By 2007 he will have served almost 20 years of his 203-year sentence, with no possibility of parole.

SOME IN EGEGIK argue that because of the alcohol, domestic abuse, and petty crime all the young people will eventually leave the village. I once met a woman who had spent a winter in Egegik, working as a schoolteacher. When I asked about her experience she told me the women were jealous of her, and the men were predatory. She felt like she was going to be raped whenever she went outside, so she spent the entire winter locked in her home when not at school, trying to stay warm, and safe. She didn't make friends and rarely had a conversation.

I told her how the place feels in the summer. I agreed that it is rough, but insisted that there are also interesting people there. And that the nature is extraordinary. I spoke of Sharon and the loft in the cabin and all the good times we have there. I told her about Carl, and his in-laws, the Deighs.

"I wouldn't go back there for a million dollars," she replied.

FISH AND GAME

WHEN I WAS YOUNG I FISHED FOR SALMON, ONE AT A TIME. I would amble down to the shores of the Sacramento River, with a pole in hand. Some of my friends had a canoe and we would paddle out to a hole and set anchor. The day would pass idly, us talking, and occasionally we would catch a fish. They were a thrill to catch, fighting me every inch of their eventual journey to the bottom of the canoe. Most got away either because I hadn't set the hook in time, or maybe because I didn't have the heart to reel them in. That's not to say I didn't love the taste of fresh salmon. I did, and still do.

Back then fish were not something I worried about, thought about, argued about, or dreamed about. That didn't happen until I began to catch salmon by the pound, thousands of them. In Egegik I have spent hours imagining the journey a single salmon takes to fulfill its final mission. I imagine the journey is the same no matter the river, whether in the Egegik or the Naknek, Kvichak, Nushagak, Togiak, or the Ugashik.

Or even the Sacramento.

The Sacramento River has two yearly runs of kings, which come once in the summer and then again in the winter. It's one of the most studied runs in the salmon field. That it has survived is a bit of a scientific and biological wonder. Imagine the journey. The fish must peel off the ball of salmon spinning counterclockwise in the Pacific at just the right time and the right place to hit the narrow opening of the San Francisco Bay. Under the Golden Gate Bridge, avoiding the noise and distractions of shipping lanes and cruise boats, they navigate the delta, a maze of channels that weave through dikes and farming communities for almost a hundred miles. Like all salmon running to spawn, they have only one path to their place of birth. Upriver they encounter an agricultural minefield of rice plantations, with dozens of false channels emptying into the river. Here the farmers suck the river and spit their waste back into it. At the same time human traffic is calming down. Other than the occasional obnoxious Jet Ski, most river visitors are sport fishermen limited to two fish per day for set dates. At this point there is hope.

Eventually the salmon pass long stretches of river where few people visit. This is a part of the Sacramento that, like the Egegik, seduces the human eye into thinking nothing has changed for thousands of years. Giant California oak trees line the shore and strangling vines climb their trunks. Occasionally there is a rope swing hanging from a limb, as if Huck Finn himself were about to take a ride. The salmon have made it. Almost.

And then they run headfirst into the fish killer. A dam. The first of two. The Red Bluff diversion dam is small and there is a way around it. Some take the detour, while others have already peeled off into their creeks, their noses homed in on some sediment or silt that comes only from that stream. Streams like Deer Creek, Antelope Creek, Big Chico Creek, or Butte Creek. Others keep racing full speed upriver. I wonder if they know the end is near. Do they know they've missed their exit? For ahead is the mother of all dams, Shasta Dam, the man-made headwaters of the Sacramento River. For fish and man alike it is a concrete wall with no ladder to climb. If a salmon hasn't found its feeder stream by this time it will just swim around for a while, move from deep pool to deep pool, and then eventually die.

Every time I find myself on the Sacramento River, with eagles perched on giant oaks, and blue herons skimming the shore like stealth jet fighters, I think of how the ancient tribes of North America practiced the art of killing with grace. They gave respect and thanks to that which they killed. They didn't do it because history would look upon them as benevolent and write stories of their enlightened environmental forethought. They did it out of fear and gratitude. Fear they would anger the gods by taking too many animals, and gratitude that the animals laid down their lives for them. The belief was that if they kept taking only what they needed, then the equation would keep working.

And yet I fish commercially and slaughter thousands. Over my four summers in Egegik our crew has brought in more than 250,000 pounds of salmon, or approximately 40,000 fish. I justify it by telling myself no fish is wasted; we are feeding our species, not something to take lightly. Still, each day I find one moment, no matter how tired I am or how much slime of their guts I have in my hair or on my body, to stare into their oval black eyes. Their mouths gasp for their last breath, and I feel the weight of guilt.

Every June and July, over the last forty years, upward of ten million sockeye swim into the Egegik River system. The escapement set by Fish and Game hovers between one and two million—the number they want to make it upriver to spawn. The rest, somewhere between seven and nine million, are what account for the commercial run. Of course seals, whales, and bears take their cut and various other species take the scraps. Sportfishing takes a tiny portion, and there is some loss of stray fish that never make it to their spawning grounds.

A salmon spawning ground is a frenzy of activity and something everyone should see at least once in life. By the thousands the females lay eggs in several redds—small beds of pebbles and sand in the shallows—and the males, always close behind, sidle up and blast out a film of milky fertilizer. The male shivers ecstatically at this point. In a few weeks their decaying bodies begin mulching into the water. I've seen salmon still spawning near death, their skin brilliant red but so thin I can see through it. Finally, after spawning, they die. The fish, now rotten from the journey, turn white and sit like ghost fish at the bottom of the streams. But this is not the end. There is a survival instinct working even in their death: their decaying bodies will be the first food the newborn salmon fry will eat upon being hatched.

This is on my mind much of the time as I stare upriver toward the volcano that towers over Lake Becharof, headwaters of the Egegik. I tell myself, if it is true—that the fish will die no matter what—then there is only one remaining factor in finding a justification for the mass slaughter that happens every summer by the fishing fleets of Bristol Bay: making sure that enough salmon get upriver to spawn, ensuring that this cycle will go on and on. If the ancient ones knew to give thanks and kill with respect and fear, then the modern method of sanctifying killing is to plan legitimately for the continuation of the species.

To protect the unborn salmon and the health of the salmon run we need something many people in Alaska would love to get rid of: the government.

Like everyone in camp I listen to the voice of the government on the radio, as the Alaska Department of Fish and Game announces the openings for fishing. I often wonder how this system works. Is there a large staff studying the salmon from satellite images? Do they have lasers in the river that count the fish? How do they make their calculations for how many fish will make it to the spawning grounds? And why do they close us on some days and let us fish on others? To find out, I arrange an interview with the Eastside Bristol Bay Office of Fish and Game, located in King Salmon.

After flying the fifty miles from Egegik, I am early for the interview. From the airport I can see the Fish and Game office, a few blocks away. I grab a newspaper and step into Eddy's, a local bar. There I keep my head down, read the paper, and have my yearly hamburger in King Salmon. I've been in Sharon's cabin for five weeks. It's a place I have come to appreciate, but there are downsides. There is the yelling of Dave Hart; the fighting between Sharon and her son; the endless drop-ins of the Egegik locals all day long, and then later the drunks, looking for the late-night party. As much as I love the chaos I am grateful every time I leave. Plus my hands are claws by now, my fingers so swollen from the work they act as one mitten-shaped unit. When I pick up a glass I use my hand like a hook, pulling the glass to my face. My liver is full of whiskey and pain pills and I need a cold beer and a burger to right myself. I am fished out and don't want to engage in any conversation. If I hear someone talking about how to fix an engine shaft or cork someone off I will either walk away or scream out loud. It is time to return to the other life, the one where screwing one another out of fish doesn't exist. I eat in silence, happy to be alive.

A father and son sit down across from me. Immediately they have the waitress turn on the television. NASCAR. They are fishermen. I can tell by the blank stare in their eyes and their reluctance to speak. If I were somewhere else I'd be inclined to think their stare is the glazed face of the modern American overdosed on the computer and television. But not here. I know the look on their faces; I've seen it in the mirror. They are tired, so tired that speaking has been reduced to nods and one-word replies.

The cars zoom around and around the track and I sit and finish my burger and beer. I read the paper, including the classifieds. There are a few permits for sale, $75,000 a pop. A few boats also. No, thank you. I leave a hefty tip and walk to the bank and deposit one check for $1,590 and another one for $2,300. I also have $500 in my pocket and a hundred pounds of filleted salmon in coolers in the airport freezer. Life is rich for those who need little.

I walk down the dirt road, swatting away mosquitoes, and enter the one-story building with an SUV parked next to it stenciled with STATE TROOPER. I tell the receptionist I have an appointment with

Slim Morstad, the Naknek/Kvichak District Area Management Biologist. She points behind me to a lanky man leaning back in a cheap office chair. He's wearing an oversized sweatshirt, Levi's, and tennis shoes.

"The most important thing is that there are no dams," says Slim, when I ask why the salmon run in Bristol Bay is healthy year after year.

I know his voice immediately; he's the voice on KDLG 90.1, on the radio announcing our openings all season. Each fishing district has a different announcement regarding their openings or closings, depending on fish count and tides. Some years all the rivers are in sync and other years one river hits big and another river has a small run. Guessing the salmon run prior to the season is a result of a statistical probability, but there are no guarantees. Slim has agreed to this interview to discuss salmon and the mystery that surrounds them. And of course why fishermen hate him.

"Fishermen all have a theory," he begins, his raspy voice low and even. "The longer they sit around not fishing the more theories they have. Usually it's about how some other district is stealing their fish. Or the belugas or the Japanese fleets."

I tell him I heard one fisherman this morning in the airport telling his crew that Fish and Game was not doing its job and that it was favoring Egegik over Naknek.

"How did he put it, exactly?"

"Well, he said, 'Those cocksucking-sonsabitches in Egegik are stealing our fish,' " I say.

Slim lets a small smile grace his lips.

"Listen. The only objective I have is escapement. Period. If a river doesn't get its escapement I have no problem shutting it down as long as it needs to get the numbers right. All season, if need be. No matter how belligerent a fisherman is in the bar, blaming us for everything, they know I'm right. No escapement means no healthy run in the future. After we get the escapement math correct then we try and allocate the harvest fairly."

Slim spends most of his time studying the data. There are scales to be tested and researched. These small flakes from the exterior of a single fish tell him where the fish was born and where it is supposed to be spawning. Fish and Game doesn't try to get the fish to go into the correct river—that is nature's job—but it does have ways of encouraging fish to find their stream. This is done by not letting drift boats venture too far into the bay from their river. This way the fish in the bay can poke their noses into various river systems in hopes of finding their home stream, while not being trapped by a net.

Slim and several other biologists study the size of the fish, categorizing them by their weight and length, which then tells them how long they have been in the ocean. Most of the data are collected during the precious few months when salmon return from the ocean to spawn. The rest of the year he studies the information in hopes of predicting how big or small the next season will be.

Sockeye salmon run in three- to five-year cycles, meaning that from the time they hatch in the lake systems they live between three and five years, depending on how long the smolt—the small stage of the fish—lives in the lake system. Humpies, or pinks, live only two years and instead of being picky about the stream they are born in they tend to flood a river system in massive numbers. Their numbers are staggering. In one season 100 million fish may hit several rivers, overwhelming the rivers' capacity to sustain so much spawning. Everyone knows when the pinks are back. The stench of death is unforgettable. And due to their soft meat and bland taste, pinks are not eaten as fillets. Instead most are canned.

Dogs, or chum salmon, are big fish that follow the sockeye upriver. They are called dogs because as they travel from salt water to fresh water their facial features turn gruesome. Their teeth grow to look like canines and their heads peel back, exposing their skulls. Fishermen throw them back or hide them in their bins for extra weight. Again, not a fish most people eat, although dog mushers of inland Alaska have historically used chum to feed their sled dogs.

Silvers, or coho salmon, are coveted by sport anglers for their fight. They arrive in Alaskan rivers about one month after the sockeye have come and gone. Silvers can be twice the size of sockeye and put up twice the fight. Unlike sockeye, who don't take to bait or lures while spawning, silvers lash out at the bait, as if punishing the lure for getting in their way.

Then there are the chinooks, also known as king salmon. These are the prizefighters of the salmon world. Known to grow up to 100 pounds in Alaskan waters, the kings are the ultimate sport fish and loved by Natives for their extra-fat bellies, which are good for smoking. Chinooks have the longest life span of all the salmon. Occasionally they last in the ocean up to eight years before coming home to spawn. Like the sockeyes, kings head for the streams where they were born.

"All the salmon runs in the Lower Forty-eight are either gone or seriously diminished," says Slim. "The Atlantic was decimated years ago by commercial fishing, but this is the most successful run in the world, and that's because management of this run started long ago and we have maintained vigilance in seeing the mission through. The fish are what matter, not the fishery."

During the season a team of fish counters camps near the headwaters of the Egegik River, just below Lake Becharof. There stands a steel tower, lashed down by cables and ropes. On top of the tower is a platform, where all season long young biology students count how many fish have passed upriver, using handheld counters. One season I went upriver with Dave Hart and visited the students. They all wore polarized sunglasses and had a casual air about them. They seemed a bit bored and ready to go home, even though the run hadn't started yet.

"I wonder if they even know," said Dave, shaking his head as he drove downriver, back to the Egegik fishing district. He was referring to the impact the students with fancy sunglasses had on the livelihood of so many fishermen.

But there are more soldiers at work. Downriver from the students, close to the upriver boundary of the Egegik fishery, two experienced biologists spend every day of the season set netting for fish before the opener. They keep their nets in the water for a short time, usually ten minutes. Using a mathematical extrapolation, they then calculate how many fish are in the river. Slim uses both pieces of information, from the tower and the two biologists, and data gathered in the bay to estimate how many fish are in the river, and how many have gone upriver, past the nets. Finally he announces on the radio when the opening, if there is one, will be.

I ask Slim if he ever feels the wrath of the fishermen.

"Not really," he says, his stoic facial expression unchanged.

I wait, wondering if he really doesn't ever ponder being stabbed by angry drunken fishermen.

Slim uncrosses his legs and leans his elbows on his knees. "I will tell you this. I don't go to Eddy's during the season," he says, referring to the bar I have just come from. "I wait until they all go home."

BECHAROF LODGE

Becharof Lodge is a complex of yellow buildings situated directly across the river from Egegik. Built originally by the Bartlett family, the compound has been in existence for two generations. As a family compound, it was the place where the parents raised their sons, but when the sons were old enough to begin operating their own commercial fishing outfits in the early 1980s, the compound morphed into a party haven for fishermen. Soon the lodge had a bar, plenty of shag carpet, and on occasion, exotic dancers flown in from The Great Alaskan Bush Company, the world-renowned strip club in Anchorage.

The Bartlett family compound began to unravel as a gathering spot when one of the sons took the family plane, some money, and a hooker, and flew away.

Soon after, another of the sons was found guilty of fish molestation, which is defined by the Alaskan fishing regulation books as "molesting or impeding spawning or the natural movement of fish contrary to lawful methods and means of sport fishing." He was fined and banned for several years from commercial fishing in Bristol Bay.

For almost a decade the compound remained idle until Phil Horton, a master guide from Washington, renamed it the Becharof Lodge and opened it up as a hunting and fishing establishment.

I first hear of the lodge from Misty, a Native girl who was once arrested while hiding there. Misty has a large, friendly smile and the round face of many Natives in Alaska. Her left leg has a scar that runs from her foot to her kneecap, part of it digging into her muscle, the result of an ATV accident a few years back.

Misty stops by Sharon's cabin throughout the season, always a welcome sight.

"I really love that girl," says Sharon. "She has the brightest spirit. But I'm pretty sure when she was younger she broke into my cabin one winter. No harm done. I never said anything and over the years we've just become friends."

Misty can be hard to follow when she has had a few drinks but over the course of listening to her story a few times I piece together the details of the time the state trooper caught her hiding at the Bartletts'.

It began with a court order. She was wanted for drunk and disorderly conduct, but when the trooper arrived by plane from King Salmon she boated across the river to hide. The state trooper got back on his plane and flew across the river. When he located Misty she walked out of a cabin, naked except for a thin towel around her privates. She told the trooper that her clothes were in the dryer and asked if he could wait until they were done, seeing she was going to jail and she had no other clothes. He agreed. She then produced a case of beer and asked if she could have one, while waiting. He agreed. By the time the clothes were dry Misty had consumed fifteen beers and was now good and drunk, ready for her own personal flight to the King Salmon jail.

"He asked me if he needed to put the handcuffs on me during the flight, as if I was dangerous," explained Misty. "I said fuck no. I'm not going to attack you 'cause I don't know how to fly."

Usually, like most fishermen, I am too tired at the end of the season to imagine one more day in Egegik, but at the end of my third season, at Carl's urging, I decide to stay at the Becharof Lodge for several weeks. By this time Sharon is long gone, her cabin boarded up for winter. The sockeye are spawning upriver, and the silvers are running the river now.

I have two good reasons to stay in Alaska. First, my girlfriend, Leigh, is working as a park ranger less than eighty miles from Egegik, and won't be finished with her contract until the middle of September. Second, I have always had a curiosity to see Egegik when the fishing fleet is gone. It also doesn't hurt that Carl has a vested interest in the lodge, where he works as a sportfishing guide until September. Then in the fall he will pick up his gun and become a moose- and bear-hunting guide.

Knowing that I can operate a video camera, Carl proposes that I make a promotional video in exchange for free room and board. My plan is to spend some restful weeks here gathering footage, and then visit Leigh in Katmai National Park.

Lounging around at the lodge has its appeal for a few days, but I find the transition from working hard to not working at all intolerable. Within a few days of watching Carl head upriver with his clients, I ask him if I can help with lodge chores. He quickly accepts and that's how I become a lodge monkey, making daily runs across the river to buy beer, engine parts, food, and gasoline. On trips upriver I shoot footage for the promotional video but also bait the clients' hooks and fillet their catches. Later, back at the lodge, Carl and I put out a commercial net to fill the freezer with fish, just to make sure the clients who didn't catch their limits have something to take home.

Most of the lodges in this part of Alaska are upscale, with names like Rainbow Lodge or Brooks Lodge, charging anywhere between $3,000 and $6,000 for a week of the "best fishing in the world." Becharof Lodge has a different angle; it advertises itself as the place to get away from the overcrowded fancy lodges where they pamper the clients with bottled wine and cloth napkins. Here at Becharof, the paying customer—$2,500 for five days—can expect fewer people but more fishing. In many ways the place is a "blue-collar" lodge and it has its share of problems.

Here on the far side of the river, the mud prohibits easy access to the boats. The electricity is unstable, running off a generator that has a mind of its own, quitting whenever it gets a bit tired. And the most obvious problem is that no one can pronounce the name: Becharof, pronounced B'SHARE-off, was the name of a Russian who stumbled across the lake upriver, the headwaters of the Egegik.

For all the thousands of commercial fishermen who have passed through Egegik over the decades, only a handful have ever made the trip to the lake thirty-two miles upriver. Most never even make it to the commercial fishing boundary, a round orange marker located a little more than a mile upriver from Sharon's site. This marker denotes the imaginary LORAN (Long Range Navigation System) line stretching across the river beyond which no commercial net can touch the water. Every year a captain falls asleep during a night opening or gets lost in the fog, and his boat drifts over. His only hope of avoiding a stiff fine is to bring his net in, or get back across the line before a Fish and Game plane or boat patrol spots him.

After years of asking people why they don't go upriver to the lake I have reduced the excuse to one common answer, "Been wanting to do that, but at the end of the season I just want to get out of here."

One morning Carl and I load up the boat for the trip upriver. We have extra gasoline, an ice chest with sandwiches, and plenty of drinking water. A six-pack of beer is hidden under the steering unit. Two clients ride with us. Four ride with Phil, the lodge manager. Phil has a habit of drinking early and continuing on through the day. This becomes all too obvious when he stumbles through the lodge dining room in time for dinner. With a whiskey red nose and a disposition similar to that of a cornered badger, Phil is a man who loves to hate his life.

Most clients who come to Alaskan fishing lodges are decked out in overpriced gear from companies such as Orvis and L.L.Bean. The ones in our boats this morning are dressed just so. Carl and I both wear the same waders we donned a few weeks earlier when we caught 70,000 pounds of fish. It's worth noting that Carl does not speak of our commercial fishing experience with these clients, with Phil, or anyone else. He will answer questions about commercial fishing, but he will not bring it up. Besides having a sense of humility when it comes to his own work, he also has a strong urge to live in the moment and not be concerned with what happened yesterday.

After ten minutes of heading upriver, the channel becomes a secret highway to a place very few have ever seen. Up here a person can spend a week and never see or hear evidence of another human being. There are no roads here. No dams. No signs. If you get badly injured, you will most likely die. If help arrives too late, probably by helicopter or floatplane, there is a good chance the bears and wolves, maybe even the foxes and coyotes, will have already feasted on your remains. All in all not a bad way to go, being a meal for the animals.

The ride to the rapids takes one hour in a jet boat and is best done during high tide, when the ocean pushes up over the shallow tundra basin and floods twenty miles or more upriver. The journey can be done at low tide but the risk of hitting rocks and sandbars increases dramatically.

Early in the trip we turn a corner to see a grizzly on the shore, eating a salmon. It hears the engine and scampers off, hiding in the thickets. Over there, two bald eagles flap their large wings, missing the surface of the river by inches. Ducks, swallows, and giant blue herons dot the shoreline, all feasting on what the river gives them. Soon they will be flying south, away from the cold. Hunkered in the tundra grass, berries have begun to ripen: salmonberries, blackberries, blueberries, raspberries, and crowberries. Hidden among the tall grass is the monkshood flower, with its seductive purple blossom, full of poison.

Carl tells me of the time when his best friend, Pete, the one who died young, took him upriver to hunt for caribou, with a man named Ben.

"We went to the lake," says Carl as he points at various landmarks along the way. "We spotted some caribou swimming and all of a sudden Ben says to Pete, 'Pull alongside.' They grab one of the caribou by the horns on his head, hold it against the side of the boat, and break its neck. When we get to shore Pete is laughing and tells me, 'We don't waste bullets on caribou.' "

EVENTUALLY THE river splits into separate channels, surrounded by tall reeds. The river narrows for a mile, but opens up again, leading to what is known as the first lagoon. This is the first place salmon stop and mill about, not so much lost as taking a moment to pause, probably because of the size of the lagoon, which resembles a large, shallow lake. More than halfway across Carl slows down and points under the boat: a preserved caribou lies on the bottom of the river. No one knows how or when the animal died, but because of the near freezing temperature of the water the body is intact, the hair and horns in perfect shape, eyes wide open.

After the lagoons we stop at Carl's favorite hole, an area of calm water just before the rapids. The clients get out and quickly line up along the riverbank. Carl and Phil rig the poles and begin everyone on a barbless Mepps lure. On each cast a fish hits on the lure, but no one can set the hook and land one.

After an hour of almost catching fish, the clients look to Carl to solve their collective problem. They see the fish, they feel them nudging their lines, but they can't catch them. While barbless hooks are the preferred, and often only, way to fish salmon in Alaska, they can make catching fish tough for the less experienced fishermen.

We get back in the boat and begin the final leg of the journey, up through the rapids. By now the salmon can probably sense their final destination: Becharof Lake.

Becharof is Alaska's second-largest lake, measuring thirty miles long and fifteen miles across. From the sky it has an oval shape, except for a large lagoon that forms its southwest side. The volcano Mount Peulik rises up, together with surrounding mountains, and forms a wall of rock shielding the lake from leaking into the Pacific Ocean, which lies only several miles away to the south.

We are almost guaranteed to see no one at this lake, giving the place an air of possibility, as if a lumbering mammoth could suddenly appear over the shore's horizon and make its way to the water for a drink.

At a hole just before the lake Carl switches the clients to salmon eggs, many taken from fish Sharon and I filleted for our home packs.

In a few minutes everyone is catching fish, and they don't stop for hours. Besides salmon, the Dolly Varden and Alaskan char populate this river, as do rainbow trout. It does not happen on this day, but it is not unheard of to catch all five species of anadromous salmon that reside on this side of the Pacific—sockeye, coho, king, chum, and humpies. There are two other anadromous salmon—fish that return from the sea to the rivers in order to breed—in the Pacific family, the masu and amago, but they are found only in Asia.

As the clients fish, their yelps of excitement audible over the roar of the river, Carl pulls out his small digital camera and flicks through the memory until he stops on a single image.

"I only saw it because I was going so slow over the rapids that I had a chance to look down," says Carl, showing me the image.

It is a picture of a giant walrus, his tusks wedged in between some boulders, presumably to help him stay underwater in one position. Only the small ripples of water on the left and right of the photo remind the viewer that this was in the water, probably twenty feet below the surface. The clarity of the image is startling.

"Can you believe it? All the way up here."

In fact walrus, like beluga whales, are known to chase salmon with a passion, even thirty miles upriver, another example of how salmon have evolved into the nutritional backbone of the entire region.

BY THE END of the day the clients have caught fifty silvers and kept thirty, the legal limit for six people.

During these trips upriver I find a conflict developing inside me in relation to sportfishing: I have lost my desire to do it.

I grew up fishing—racing to the river with wide eyes in anticipation of that first cast, and the setting of the hook when the fish gently tugs at the end of the line. As a kid I dreamed of the summer trips to Stuart Fork in the Trinity Alps to catch rainbow trout. Later, as a teenager, I would light out on the weekend, before the sun had a chance to rise, to catch trout on Indian Creek, Mill Creek, or Antelope Creek.

My guess is that seeing so many fish die during the commercial season has ruined my desire. Fishing has become something I do for a job, not for pleasure.

And sportfishing has changed over the past decades. In Alaska, most sportfishing is catch-and-release, with a small daily limit on how many fish a person can keep, depending on the river and species of fish. This method of control is in place throughout the prime fishing rivers in North America, and for good reason. They are crowded with people, not unlike commercial fishermen, who if given a chance will take more than they should. Still, the very idea of catch-and-release is a compromise I am unwilling to make. There is no sport left for me there. I either fish to eat or I don't fish at all. Hooking fish, playing with them on the line, and then releasing them, injured, into the water, especially those with a gaping hole in the lip, doesn't feel natural. No, I would rather throw in a line, sans hook, and just spend the day walking the bank, dreaming of the fish darting with lightning speed in the shadows of the water.

• • •

The long windows in the dining room at the Becharof Lodge offer an omnipresent view of the river. At high tide the river is a mighty force, the water filling the channel from bank to bank, the ocean current bucking up against the downward flow of the river, creating rollers. During the season I dread these rollers, which are responsible for making my body sway uncontrollably for hours after being back on land.

Several hours later the river is at low tide. Now the channel is nothing more than a floodplain of mud, the water reduced to a few channels etched through the earth, flowing back toward the bay. At low tide the smells of rotting salmon and boat scum fill the air. Also during the ebb, the ocean tides and the river flow out together, creating a deceptive riptide, often the most dangerous and violent time on the river. A simple mistake during the rip can get a captain and his crew quickly killed. If the skiff gets stuck sideways the current will flip the boat. If a lead line catches on a prop the sinking will be fast and almost impossible to survive.

I contemplate these catastrophic possibilities as I look out from the lodge windows, my mind still reeling from all the things that could have gone very wrong this season. And the seasons before that. Last year a lead line snagged on a boat owned by a family that fishes near Sharon's sites. The boat went sideways and flipped over. Two young men fell in and only one came out. They found the body of the other man six days later, downstream.

At the bottom of low tide, just before it begins to flood again, the river calms down, showing its softer side. Sandbars hidden by high tides now become large landmasses. In some places sandbars five feet tall suddenly appear, creating deep channels for the boats. Upriver low tide is even more dramatic, where the banks can tower ten to fifteen feet above the water's surface. These same banks disappear to the naked eye during high tide, when the boats cruise along the surface, offering no hint at what lies beneath.

• • •

On days I don't go upriver, I take the ATV out on the tundra. As the weeks pass I realize the best part of living on the far side of the river is how the landscape comes alive once the buzz of the fishing fleet is gone. The flowers seem brighter, the birds louder, and the grizzlies more abundant. Of course none of this is true, but for possibly the first time since coming to Alaska I am able to stop and watch. And listen.

The swallows come and go on the bank, checking on their newborns. The sandhill cranes squawk as they launch from the shore out over the river, their bodies long and elegant, their voices loud and gravelly. I will see these same birds in two months' time, back in southern Arizona, as they come south for the winter.

On the ground, Alaska's wild plants come to life in late summer; the orach, chickweed, and wormwood all make appearances, as does the highly coveted lingonberry, which the Natives collect for their winter berry stash.

As I drive out to the King Salmon River over the spongy tundra below, it never occurs to me that there will be any traffic. This part of Alaska is a vast, open plateau of tundra crisscrossed by small streams, alder thickets, and shallow ponds. During the winter, snowmobiles are the preferred mode of transportation. A person who knows his way can travel to the villages of King Salmon or Ugashik and beyond, as long as the rivers are frozen solid.

But for now, there is no one out here and won't be until the first snow, when a handful of hunters will fly in for moose and bear season.

On my ATV I crawl over small hills until I reach the cliff, some hundred feet above the confluence of the King Salmon and Egegik rivers. I walk to the edge; my feet struggle to find their footing as I traverse deep pits covered with thick tundra grass. These pits are original Egegik campsites, dating back hundreds of years, if not thousands. How many years exactly? No one knows because people rarely come to study this part of the world. It is also unusual for a local person to remember past his grandfather's stories. For all the thousands of fishermen and local Natives who have traversed this land there is very little evidence of their presence here. Everything here eventually disappears, swallowed by the land, including the people and their own history.

I take a deep breath. It's a strange feeling knowing that no one I know—outside of the few I've met here—will ever visit this place. They may travel to the Great Wall of China or the peak of Kilimanjaro, but they will not step foot in this place. I am overwhelmed with gratitude that this place does not speak my language or keep time by a clock. Then a feeling takes shape, washing over me, shortening my breath. I suddenly realize I am grateful to be standing in a place that does not welcome me.

Sometime near the end of August I begin to sour on Egegik, or, more specifically, the lodge.

The VH radio hisses in the background and the hum of the generator vibrates through the walls. The wind hasn't stopped for days and all the birds and grizzlies are out of sight, hunkered down, waiting for a break in the weather. The wind beats down on the vinyl siding and metal roofs with tropical-storm force, bringing with it a permanent whistle. No matter where I go in the lodge the wind whistles through the walls, windows, and cracks.

So I sit in the window staring at the river, counting the days until I leave.

Back when Carl asked me to spend several weeks at the lodge I didn't hesitate. And as the weeks have passed only the wilderness of the area has lifted my spirits and made me want to stay longer. The more the land has been indifferent to me, the more I appreciate its brutal honesty. At the same time I can't wait to leave and, strangely, the reason I begin to plan my escape has nothing to do with the approaching winter and the elements but rather with another beast that roams the barren landscape of Egegik: man.

"So you're a writer, huh?"

John, the man pacing the room for the last thirty minutes, is a Kiwi. As usual he is drinking beer, opening a fresh can of Budweiser the moment he finishes the one in his hand. For the past month John has been stopping by the lodge to visit, each time more belligerent than the last. Whatever the conversation, it always comes back to this: "So you're a writer?"

I tell him I am a writer as well as a fisherman. He doesn't hear that last part, instead fixating on me being a writer.

"When was Walt Whitman born?"

I tell him I have no idea.

"And you call yourself a writer?" He laughs and blurts out a date he says Whitman was born.

"How many books did Mark Twain write?"

Again I admit I have no idea, but tell him that Huckleberry Finn is one of the great characters in Western literature.

The man shifts his weight from one large foot to the other. His steel-toed boots are not lost on me. "Think you are better, do you? Just because I'm driving a forklift you think you're better?"

"I've spent time driving a forklift as well."

"Oh, so now you can do my job too?"

John has a job half a mile down the river. He and another man have been hired to tear down the oldest cannery still standing in Egegik, which has been inoperative for decades. The thick lumber is in high demand for the exploding housing market in the Lower 48.

"All of it is being shipped to some fancy housing estates in Montana," John explains, spittle spewing out of his mouth. Six feet three, with a blond mustache and thick mop of hair, he has expressive features and a rich vocabulary. He likes to talk and drink, and the more he drinks the more he talks. Usually he drinks until he passes out, which isn't the problem. It's the anger he builds up just before crashing.

I suppose it was inevitable that I should begin to feel the impending darkness of winter, accompanied by a darkness in my state of mind. In the beginning the landscape and the rush of going upriver were enough to distract from the reality of the dysfunctional people around me. That, along with the fact that I came to the lodge with the same attitude I have when I enter any job: that if I work hard enough I can earn the respect of those who have been there longer.

After several weeks I understand better with each passing day that in the bush one often develops a vicious survivalist attitude not unlike, I suspect, what inmates encounter in prison. And though I have tried to prove myself through work and river knowledge, by the end of my stay I have become the outsider, the mark at the poker table; and after a while I feel almost like Piggy in *Lord of the Flies*, being readied for the feast.

Most nights the conversation over beer and whiskey veers toward talk of guns and fishing reels and which bait someone used on some river in Canada, or how a guy shot a deer by sitting in a tree for three hours. When these conversations erupt into drunken shouting matches I tend to slip back to my cabin with two or three beers, where the wind crushes down on the roof like a giant hammer.

My silence, however, is not appreciated. My crime is simple: I am a writer. It is not until late in my stay that I realize I have been labeled as an intellectual trespasser in a world where only the loud and reckless are welcome. My experience in Egegik all these years does not matter. What matters is that I don't play the part of the great outdoorsman.

At the moment Phil is in the kitchen, listening to Kiwi John rant on, calling me a writer in the same tone as if accusing me of stealing his last dollar. Drunk, Phil walks out of the kitchen, bumping into the couch twice as he approaches me at the window. He gets close, sticks out his pointer finger, and says, "You are just being superfluous," which he pronounces soo-perfa-liss.

"Phil, I'm a writer. I'm broke. That's why I fish with Carl."

"See, that's what I mean. Soo-perfa-liss!" he screams at John with great glee.

I try to laugh along, not knowing what other line of defense to take. Still, I know Phil and John believe writers are a useless bunch, unless of course they are writing heroic stories about them.

I escape to the bathroom only to find John following me down the hall.

"So what are you writing about?" he yells after me. I turn and see his eyebrows wrinkled, exposing what appears to be curiosity.

"Right now, nothing," I reply. "Haven't been able to write shit, but I hope to write something about this place."

"Well, maybe it will have something about me," he suggests with some optimism, his arm propped against the wall for support.

"I'm sure it will," I say, and close the door.

KATMAI

GOING HOME CAN WAIT. I WALK THROUGH THE FOREST AND celebrate its geographical difference from Egegik. A thick fog seeps through the tall pine trees. Steep mountains rise in all directions capped by glaciers, which melt thousand-year-old ice into the rivers below. This is the geographical dreamscape that comes to mind when people say they are going to visit Alaska. It is immediately enchanting, and all I can think of is how glad I am to be out of Egegik.

Other than my girlfriend, there is no one else on the trail. I left Becharof Lodge two days ago, deciding it was time I visit Leigh here in the place she calls her office: the five million acres of Katmai National Park, where she works as a park ranger.

Needing relief, I step off the trail and piss in the woods.

"Bill," Leigh whispers.

Finishing up the task at hand, I turn my head and see her face. She is not terrified. That would be overstating it. She is concerned. Very concerned.

"Right in front of you," she says quietly, careful not to make any sudden movements.

And there he is, no more than thirty feet in front of me: a massive grizzly bear, perhaps eight hundred pounds, lying stretched out on his back. Because of his dark coat the creature looks like a downed tree or a decaying log. I freeze in place and Leigh—used to these close encounters—whispers for me to back up slowly. The rules of engagement when encountering a grizzly race through my mind: Back up, talk softly, don't act weak, yet never threaten, and never, ever, turn your back on the bear. If eye contact is made, avert eyes, and offer the bear your nonthreatening profile. As for playing dead, that is the last resort and should be done only if the bear makes physical contact. Before that, you invite a mauling.

Slowly we make it back to the trail, just in time to witness the bear lift his head and stare in our direction. Frozen with fear, and yet thrilled to be so close to one of the most feared beasts in the world, I wait for the next move.

Then the bear looks away, scratches his belly, and lays his head back down on a pillow of leaves.

"He just ate. He's fine," says Leigh, as we continue down the path toward the stream. In the distance there is a loud roar, a noise I've never heard before.

"We should get to a platform," says Leigh, referring to one of the two platforms in the park that allow humans to observe bears from elevated positions behind the safety of steel-mesh gates.

"What in the hell was that?" I whisper as Leigh opens one of the gates and ushers us down a long wood platform high above the ground.

She explains it is most likely two males fighting over a mate, or a male grizzly chasing after a female. As we walk I notice I can still see birds flying from limb to limb but can no longer hear them. The pounding of the blood in my ears is much too loud.

• • •

LESS THAN EIGHTY miles from Egegik as the crow flies, the round-trip flight to Brooks Camp costs just shy of $500. Brooks Camp is a tiny outpost of people situated in the most bear-infested zone of Katmai National Park and sits on a two-mile stretch of land between Naknek Lake and Brooks Lake. The two lakes are connected by the Brooks River, which serves as a popular spawning ground for salmon, which in turn attracts the concentrated population of grizzly bears looking for an easy meal. During the peak of the spawning season a person may spot up to sixty individual bears in a day.

Beautiful yet disturbing, the camp is like a futuristic Jurassic Park, where the predators are on the loose and tourists are shepherded around the grounds by park rangers. Although there is no fee to enter the park, getting here is expensive, and the only place to sleep indoors is the private lodge that runs as much as $350 per night. Meals are extra. There is a level campground nearby surrounded with a somewhat undependable electric fence, a likely reason why most people opt either to stay at the lodge or to take a plane out the same day of their arrival.

When the park is crowded with tourists it can be easy to forget that not much separates the 800-pound bears from us. But this is no zoo. Other than the few vaulted platforms and a wood fence that encloses the bridge, the park is open land, and the only reason the grizzlies don't get up on the platforms or tear down the fencing is because they choose not to. Instead they ignore us, to a certain degree. The rules of walking in Brooks Camp are simple. Bears have the right-of-way. If a bear wants to walk on the path, then it's time for you to back into the forest. If he wants to sleep on the bridge then a visitor can be stuck for hours at a time until the nap is over and the bear moves on. And because bears are given the right-of-way in Katmai, rangers are specifically instructed not to interfere with the bears on the behalf of humans. One day I stood with several other people as a mother and two cubs napped on the bridge. A couple in front became irate at the ranger for not moving the bears. Finally the bears moved along, disappearing into the forest. Maybe they grew tired of the whining. Strange how some people come to Katmai to see bears in the wild and then don't like it when they are told they must wait for Mother Nature.

If, by chance, the bears do get too close to the humans, the park service bear technicians are on patrol, ready to defuse the situation. A bear once trapped Leigh in an outhouse, trying to open the door. She was saved by a technician who blasted an air horn in the bear's face.

A small but interesting fact is that I never see a bear technician with a gun. Instead most carry a large can of pepper spray, an air horn, and a loud voice for protection. They are firm in their belief—backed up by experience—that carrying a gun with the intent to deal with a grizzly is futile. They argue that grizzly bears are naturally afraid of humans and the trick is to understand how a bear acts and reacts when encountered face-to-face.

All this precaution is advertised as a way to protect humans from the bears, but when I ask rangers how they view the place, they all believe the vaulted platforms and gated bridges are to protect the bears from our intrusion into their forest. How refreshing that they understand who's actually dangerous.

Every morning for the next three weeks I wake to the sound of floatplanes coming and going, dropping off and picking up tourists, each having shelled out hundreds of dollars for a chance to see bears in their natural habitat. And I am just as excited to see the bears as any tourist flying in for the day. But luckily I am able to sleep in my girlfriend's cabin, courtesy of the U.S. government. It is all I need: a wall tent with a lofted bed, a stove, and an electric heater. And a beautiful girl. The outhouse is nearby and hot showers are just down the path. There is a weekly poker game with the lodge employees and fourteen hours of bear viewing a day.

Each day I do the same thing as every other visitor: walk down to the river and watch the bears eat, sleep, and occasionally fight. It takes only a few days to realize it is a routine that I could happily repeat on a daily basis for the rest of my foreseeable life. At night, protected from the rainstorms, Leigh and I hunker in bed and talk of returning every summer to Katmai until we are too old to walk the few miles from camp to the waterfalls. We will return just to gaze at the bears.

There is no known drug or man-made fantasy that can take a person this high. Being around bears in the wild makes all the external buzz of my life come to an unexpected halt, and I end up with no urge to be anywhere else.

To feel the grizzlies' presence fills my head and soul with an overwhelming sense of relief that they are alive and thriving, at least in this one spot. It gives me hope that the bears will find a way to outlast the human desire to mount their heads on walls and use their thick coats as rugs.

I ALSO FIND myself standing for hours near Brooks Falls, a waterfall one mile from camp, to watch salmon spawn. There, every morning, when the light is best, I venture to the falls and watch the female salmon fight each other for small patches of gravel where they lay their eggs. In Egegik we only see chrome-colored salmon, fresh from the salt water. They still have their silvery skins with taut muscles and alert eyes. At the most they have been exposed to fresh water for twenty-four hours. Up here, thirty miles from the nearest salt water, the sockeyes' skin has turned deep red. They now have large humps on their backs and their jaws protrude, making them appear violent and slightly mad. If a person didn't know better one could easily believe these were two different species of salmon, instead of the same fish a few weeks apart in its life cycle.

The bears are the draw for tourists to Brooks Camp, but down below is the true drama of Katmai: salmon fighting and struggling to lay their eggs. After she lays her eggs the female will hover above her shallow nest until her energy wanes. Within a week, or two at the most, she will die. The males will follow.

• • •

At night it is time for a drink at the bar, where park rangers, lodge workers, and visitors gather for beer, games of Scrabble, and the ancient game of mating. Occasionally I get into heated discussions with a ranger about commercial fishing, which many of them see as an industry hell-bent on destroying salmon. Most don't know it's a managed fishery with strict enforcement. They don't know that just as in Katmai, where the bears are the first priority, in Bristol Bay the health of the salmon run comes before the livelihood of the fishermen. These conversations last entire evenings, each side arguing the pros and cons of trying to balance being alive on a planet that needs more and more attention from us.

Toward the end of my stay at Katmai I go fishing. With a pole and lures. I cast just below the bridge where bear cubs often jam up the tourists as they play with the gate latches on either end. I cast into a swarm of fish. I don't really want to catch one, although if I do I will take it back to Leigh's cabin and cook a meal for two.

Being away from Egegik, but still in Alaska, reminds me why I love coming to this state. Everything still seems possible here. It's why people move to Alaska with their dreams of living off the land. There is wild game to trap, fish, and hunt. There is wood for fire. Of course very few people are able to live completely off the land. Some say there are no mountain men and women left, only a handful of part-timers, who come to the city for necessities. Still, I wonder what it would be like.

Being here in Katmai, where the pulse of life slows down, I feel much different than just a few days before in Egegik. Egegik is not a place to ease one's mind, or even relax. It is a place to work and survive. Friendships are deep, and I have made some, but on any given night the conversations tend to be about boats, fishing, money, tactics, and how best to fix a broken head gasket or replace a drive shaft. The time spent in the Becharof Lodge only reaffirmed my suspicions that living in Egegik would be a lonely existence, and I say that without even having lived there for a winter. As much as I enjoy fishing with Sharon and Carl, I also enjoy sitting in the Katmai lodge having another pint of draft beer and talking about the true value of bears and salmon. It feels sane, even therapeutic. It also makes me wonder if I will ever return to Egegik again.

Without much effort I catch a fish. It's deep red and missing a piece of its back, perhaps the result of a bear swiping at it in shallow water. I can't imagine eating this fish, and instead release the hook and set it free, laughing at myself. Given time, I will once again find joy in fishing, with a pole, on a river, with nothing more than the sound of water rolling down a canyon carved by ice.

Before flying back to the Sonoran desert for the winter, Leigh and I stay a few days in King Salmon, the way station for flights to and from Anchorage. On our last evening we drive down to the Naknek River and park on the bank. To our right sits a hundred-foot-long tender. A man stands on deck tying ropes. The boat is high and dry, waiting for the tide to come in and lift it off the ground.

Down the river are several canneries, which for almost a hundred years have been processing, canning, and shipping salmon to Europe, the Lower 48, and Asia.

My mind drifts to the cannery directly across the river from us. I know this cannery. This is the cannery that shy Kevin—who works for Earl—was referring to two years ago, as I stood in Sharon's kitchen baking a salmon. He had said, "You were dating that really cute girl."

There is nobody at the cannery now; the college students and foreign workers are long gone. The last of the crews left weeks ago, winterizing the place for seven months of below-freezing temperatures. The barracks, most of them old and run-down, have rooms for two or four people. I don't know about now, but there used to be a secret hot tub just behind the egg house, the building where women work eight-hour shifts sorting out salmon eggs for Japanese buyers. I know this place because it is where I met that girl so long ago. I was twenty-five and she was nineteen. It's strange to think that her death, fifteen years ago, ignited an odyssey that has led me back to this place once again.

I look across this river at the decaying cannery, and then back at Leigh. She and I met in a poker game in Arizona five years ago. Leigh was wearing a cowboy hat and winning all the hands. I knew I would chase her everywhere she went. And we have been coming to Alaska for three years, she to work in the wilderness, I to fish.

I feel time slowing down as the past and present collide, as if two loves, one alive, one dead, are mingling in the river before me. Thankfully, over the years, the grief has dissipated, replaced only by good memories. My gaze is no longer in the past but toward the future.

Commercial fishing leaves little room for philosophical reflections. But here, I have time to appreciate the cycle of these fish and their genetic drive to spawn no matter the obstacles. In the spring, when the eggs hatch, the smolt will eat the dead remains of their parents. When strong enough they will spend time in the lake, growing. The survivors will swim to the ocean to rejoin the giant wheel of fish constantly swirling through the Pacific Ocean, like a giant clock, turning backward, as if counting down the days of their own existence.

I am not dying, and presumably still have several decades left on my own clock, but staring at the cannery across the river I definitely feel a cycle coming to a close. I can't help but think of something Sharon said to me once: "We are just like the salmon, dying to let the new generations live."

I feel a shift inside. Maybe it is a small death. If so, I am glad. I am also glad I am not yet at my final resting place. Suddenly I feel an urgency to go home.

THE LAST SEASON

EVERY TIME I LEAVE EGEGIK I AM CONVINCED I WILL NOT return. I tell Leigh I will not go back, under any circumstances. The pain in my arms grows worse by the year, a chronic ache that will only get worse if I fish. The most money I ever made in one season was $6,000, the least around $3,000, hardly enough to justify the hardship. Between the mosquitoes, the bears, and the wind, a person is trapped inside most of the time, and Sharon's cabin, although cozy, becomes smaller every year.

When I say I'll quit I always mean it, at least until I have all winter to forget why I wanted to quit in the first place. Then each spring I begin to worry that Sharon will not call, that she has found someone new to replace me.

On the eve of my fourth season I arrive in Anchorage too late to make the connecting flight to King Salmon. I check into the Motel 6, which in the summer costs $156 for a single person in a double bed. I arrive with a limp due to a severely swollen foot, the result of a three-month-long battle with acute gout. I've had the affliction most of my adult life, and have learned to control it by moderation and medication, but sometimes the toxins get the better of me. In my luggage is a bottle of pills that are useless for stopping gout but do wonders for subsiding the swelling. I have already taken two cycles of these pills and my doctor told me while writing out the latest prescription, "These will stop the swelling but if you take these pills for too long they will kill you."

Knowing I can't fish in this condition, I reach into my duffel bag and begin what will become my third cycle of steroids.

I have only myself to blame for my predicament. In the past six months I married Leigh and bought a house—meaning I assumed a large debt, which has a way of manifesting in nightly dreams as a pack of hyenas eating me alive. I also decided to remodel the house, leading to more hyenas. For most, gout is usually related to diet. For me it has always attacked when I was either overstressed or poor, which most will agree go hand in hand.

So that's how I enter the fourth season. Happy, crippled, and ready to quit before I even begin.

CARL PICKS ME up at the airport in shorts and a sleeveless T-shirt. It's in the mid-50s, a balmy day in Egegik.

"How was your winter?" he asks, not waiting for a reply. "We have a slight problem," he says, as we drive to Sharon's house.

"Never heard you say that," I say.

"The mud is thick at our sites."

"How thick?"

"Too thick."

"Can't say I like the sound of that," I say sheepishly, my mind flashing on the image of a little girl stuck in the mud and being torn in half before she drowns.

"How's the gout?" he asks.

"If I take the pill three times a day I should be fine. But a big day and my ankle will buckle."

"It'll be an easy season," he says, pausing for my reaction.

"Yeah, until we have twenty thousand pounds to pick and we're out of time."

Coming back to Egegik always feels the same. I am drunk with the sight of eagles flying and foxes crossing the road. I wonder how long it will take to see a grizzly wandering through Sharon's backyard. At the same time I dread the yelling and bickering that come with inhabiting Sharon's world. As for the work, I always start off full of excitement to get on the water, knowing that the daily routine will soon become a monotonous, painful chore that will have me wanting to leave as soon as the season is finished.

WE SPEND THE first three days changing all the gear from 50-fathom nets (300 feet) to 25-fathom nets (150 feet). Because the mud is much deeper this year we also change to a double anchor system. Instead of using a line that reaches down from the stakes onshore, we will attach the net to an anchor and drop it near the shore. This allows us to fish near the bank but keeps us from having to walk in the mud. The other advantage of a shorter net is that it takes us out of the deep part of the channel, where the wind creates large roller waves, making pulling the net especially difficult.

The first day out we drop a sixty-six-pound anchor from the boat near the shore. But even after all the preparation to avoid the mud, we still have to get in the water to overhaul the net onto the beach and free our buoys from the mud.

My first step onshore I sink to the middle of my thigh. Stuck, I look over at Sharon, who is laughing.

"I told you it's deeper this year," she says.

It takes thirty minutes to get the net onto the beach and get the anchors in place. As usual we are the last ones to pull our net out.

"Why be first, when we can be last?" asks Carl, goading Sharon, who usually makes us late by her inability to wake up, make coffee, and get ready in time.

"Oh, now you want perfection," says Sharon as we drop the other anchor in the channel, creating a dangling net between two anchors.

All day we work in our T-shirts under a bright Alaskan sun. We bag 5,000 pounds and spend the afternoon laughing and retelling old stories, three friends who share a common experience, bonded by the water, the fish, and the work.

By the time we get to the tender it is just before midnight. The temperature is now in the low 40s and I am shivering.

Maybe that's what makes me drop an anchor on my thumb.

FROM MY DIARY, Thursday, June 29:

> My left thumb is black and blue. The back side is green, the meat tenderized. I want to avoid pushing a needle through it only because the chances of getting fish poisoning in an open wound would be much worse than this pain. Fish poisoning turns the finger into a swollen digit that oozes pus and looks as if amputation is the only solution. And my right forearm has already developed severe tendonitis, a painful stabbing sensation that limits how I pick fish. The thumb will recover with each passing day. The forearm will only get worse.

Nursing my forearm, I stop by the village store to pick up some chocolate—a key item for breaks on the water—and eggs. Kevin Deigh sits behind the counter, reading a book and still wearing his standard summer outfit: jeans, Marlboro hat, T-shirt, and jeans jacket.

"Bill Tyler," he yells at me with a giant smile. "I thought you were dead."

And so, the inside joke continues, and for the first time I feel touched that this is how he greets me. Maybe it is because he thinks of me this way: alive, and not dead.

Happy to see him, I ask what he is reading.

"German sniper in World War Two. He escapes into Russia and it's the story of how many people he killed in the war and how he escaped. Fucking cool guy. He would be a great hunter."

I take note of the topic and then move on.

"Get out of Egegik this winter?"

He stands up. "First time in three years I made it past King Salmon. All the way to Anchorage."

"And?"

"I was there for a week. Man, they got everything there. Burger King. McDonald's, I mean everything."

"Where did you stay?"

"In a hotel by the airport."

Carl tells me that Janelle, Kevin's sister, and the Deigh family are more worried than ever about Kevin. Looking at his puffy ankles and the way he walks, I can see that his feet are more swollen than the last year.

It takes a few days of being in Egegik to catch up on the village gossip. In my first few years I didn't know enough people even to care about the gossip. Now I return each year hoping that Bunny still runs the post office, that Dick Deigh is still the mayor, and that Paul Boskoffsky is still fishing. I want to see Misty alive and well and telling stories, and hear how her parents, Wendy and Henry, are doing. Scovi is impossible to avoid and over the years I have begun to seek her out, thinking that a head-on approach might deflect her wrath.

As for the gossip I want to know who left town, who died, and who went to jail.

Reggie, a Native man with a big smile and a weakness for white drugs, is in jail. In years past he always dropped by Sharon's to see if she wouldn't mind sleeping with him.

"Sharon," he would say, smiling at her. "Sharon, Sharon, Sharon."

After a few moments of laughing with Sharon he would say it again, "Sharon."

She would soon find a way to get busy doing something—washing dishes, cleaning out her wet bag, or getting involved in paperwork. Finally Reggie would leave, repeating Sharon's name over and over as he shut the door.

Reggie was arrested this winter for molesting a young teenage girl, although no one was forthcoming about the story behind his alleged crime.

Early in the season Kanut, the seventy-something Finnish captain of the *Golden Boy*, had a seizure and rammed his boat full throttle into a tender. A few of his crew, resting belowdecks at the time, were injured.

Gabe from Club Ohio didn't return this season because he was thrown in jail following last season. Something about drugs and assault.

One day, not long after I arrive, Dave Mandich, captain of the *Predator*, pulls up to Sharon's house in his four-wheel-drive truck. We shake hands and make small talk about the winter.

"Sam died," he says, referring to one of his crew. "Burned to death in his own house."

Sam. He was a Native man who came over to Sharon's cabin several times a season with Greg, Dave's other crewman. They would smoke and drink and tell stories of working on the *Predator*, one of the most productive drift boats in Egegik. Sam always stood in the corner, a smile permanently plastered on his face, smelling so badly of rotten fish he would clear a room. A kind person with a soft voice, when he spoke he had a thick Native accent, softened by a slight slur.

Once I asked about his home village, New Stuyahok, up the Wood River from Dillingham, directly across Bristol Bay from Egegik.

"I work the season and then go home and don't work for the rest of the year," he said. It was believable, because on a good year, working crew for the *Predator*, a person can easily make $20,000.

"In the winter I hunt for moose and trap, but that's it," he explained to me. Sam was living the good life and he knew it.

And then the kerosene lanterns blew over and lit his house on fire, killing him in his sleep.

I hear some bad news from Carl too. His friend Charlie from Naknek was in Egegik before the season started to repair the engines of the *Coho*, a tender tied up to the city dock. One afternoon Charlie drove his three-wheeler into the dock railing; the impact threw him over the railing and onto the ship below. The local medic pumped six pints of blood back into him before he could be flown to Anchorage for emergency surgery. It is a reminder that advanced medical care is two to three hours away. If the trauma is serious, the victim can only hope that there is someone at his side to calm him down, warm him up, and stop the loss of blood. Everyone agrees the village medic saved his life, at least for those first few hours. He had a punctured liver and a perforated lung. Carl has heard that his last meal before crashing—black beans and rice—has seeped into his lungs.

AROUND 10:00 P.M. one night Scott Olsen drops by, the local man who has a hard time fishing if he can't be the captain of his own boat. He tells us that his wife, Michelle, whom I met on Sharon's boat my first day in Egegik, is in Washington giving birth to their second baby.

"The boy will be part Native," explains Scott to Sharon. "He'll look white but underneath he'll be Native. He'll have that warrior strength."

"Sounds good to me," Sharon says, smiling in my direction and rolling her eyes, as she fries some chicken for her son.

David Hart enters the room just in time to hear Scott talk about the trouble he's having getting his boat launched.

"Just get it in the fucking water. Jesus," yells Dave, no doubt his version of tough love. He's hoping to pressure Scott to actually fish. Scott laughs, agreeing, if only he could get a certain part for the engine.

Later, Mike and Kevin Deigh open the door with a twelve-pack of beer in hand.

"Where's Carl?" asks Mike.

"Upstairs," says Sharon.

Everyone but Sharon and I go upstairs.

Sharon pours me a glass of boxed wine and one for herself. The upstairs laughter echoes down the stairs, followed by bursts of sporadic storytelling. The late-night sun approaches midnight, casting the room in an orange glow. I look at Sharon's hands.

"Had it done this winter," she says showing me her wrists, exposing small scars on each where the hand meets the arm. She had carpal tunnel surgery to relieve the pain in her forearms.

"Still hurts when I have to use the whole hand," she says, moving her hands in circular motions.

The hard part of the season is still ahead of us, and when it begins we will retreat to our private rhythms of work, pain, and sleep. But tonight we all have a light, giddy feeling of participating in something secret, away from the eyes of anyone who is not here. No one can imagine what it feels like to be here, to feel this freedom. It's something that has taken me a few years to figure out, but now it's something I feel part of. Out here it really doesn't matter if I am a screwup, a drunk, a liar, or a thief. What matters is that I do the work and operate inside the unspoken code of fishing conduct. And once a person joins this union of fishermen, forged long before things were written down on paper, he is part of a family, whether he likes it or not.

Sharon and I take our glasses and walk to the bottom of the stairs. She stops and raises her glass. "Welcome back to Egegik."

And like two kids racing for their favorite fishing hole, we bound up the stairs, anxious not to miss one more beat of laughter, or one more moment of the story.

SINKING

EVERY SEASON WE ALMOST DROWN. IT'S INEVITABLE. THERE IS always one time when everything turns against us, forcing us to use all the skills we've learned to escape. Although always on the same river, in the same boats, on a day with too many fish, no two days of fishing are ever the same. Each crisis reveals something no one ever expected.

As we begin our fourth season of fishing together, it's clear we have never fully recovered from last year, when the nets broke. The physical exhaustion that followed that day lingered for almost a week. And toward the end of the last summer Sharon admitted, "It was one of the most dangerous moments I ever experienced on water."

The three of us never doubt we can handle the nets in the worst conditions, but on peak days everyone can use an extra hand. With that in mind Sharon has hired Joe for the season, a twenty-year-old from Puget Sound whom Sharon met during her seine-fishing work in Washington over the past winter. He will get a greenhorn cut for the entire catch, but the real reason he is here is to be an extra hand on the one or two days of the peak.

So far Joe is doing well, more relaxed on a boat by far than I was my first season. The social misplacement of being a rookie is tough on him, but I tell him it was like that for me too. People come to the house and never ask his name, speak to him, or look at him. Meanwhile, I catch myself in conversation with these people, talking for hours on end about local drama, using local names and references, behavior I found annoying when I first came to Egegik. And I too ignore Joe as the season progresses, not wanting to spend the effort filling in all the blanks.

But Joe is a true boat monkey. He's comfortable on the water, can handle the nets, and is at ease in the company of loud, overbearing fishermen. He is a natural, and it doesn't take long before worrying about Joe isn't necessary at all.

Every opening Carl and I take his skiff across river, while Sharon and Joe take the other. Once on the other side we use both skiffs to drag out the nets. Then we anchor Carl's skiff and use Sharon's skiff to pick fish from both nets.

For the first few weeks it looks to be a drifters' season, the fish running hot on the outside but not hitting upriver. The escapement is normal, but Fish and Game keeps us off the river during most of the high-tide pulses, hampering our ability to fish during the best conditions. But I don't mind. It is a relaxing season and Joe's presence means we can let him do the heavy lifting. Carl is thirty-nine, I am forty, and Sharon is forty-three. Joe is twenty. The math tells us to let the young guy tweak some tendons.

Last year the peak for us was on June 26 and a complete surprise. This year the peak doesn't hit until after the Fourth of July, and we are ready. Or so we tell ourselves.

A few hours before opening, Carl and I put a third boat in the water. We will tow it across the river because it has no motor; we need it only as a mini-tender, a vessel to hold our nets and extra fish if we get overloaded.

The air has a marine odor, still thick with the scent of last night's minus tide. On our first set we bag 200 pounds, not an omen for a big day. For two hours we work the gear, hoping for a pulse of fish. Finally, at high water, when the current is going downstream and the tide is going upstream, which slows the river to a crawl, we pick one more time. Still, nothing much.

We spend the next thirty minutes onshore to pass the time. It's one of the secrets of fishing. A lot of time fishing is spent waiting. Waiting for the fish. Waiting in line at the tender. Waiting for Fish and Game to make their calculations. Waiting for your captain to make a decision.

So we wait. Meanwhile, everyone tells stories of what they want to do this winter. Carl talks about his plans for the lodge. Sharon needs to build a fence around her yard, at her home in Washington. Joe is trying to reconnect with his father, who works as a mechanic at a high-priced fishing lodge nearby. Finally we decide to pick one last time before we go back to Sharon's for lunch.

Joe and I grab the net and immediately I turn my head to the back of the boat.

"In 'em!" I yell, while using all my strength to pull the net up and over the bow.

The first net is plugged and so is the second. The peak is here. After a few hours Sharon's boat is almost full. Already we are up to our knees in fish and have little room to maneuver.

We want to keep working the nets, pushing hard, but at the same time we must manage how long each net is in the water, allowing for enough time to get them on board before midnight, when the opening ends.

"Overhaul this net, then we go back and get the first one," shouts Sharon, her hair spraying everywhere like grass in the wind. "Then we overhaul that one. That should give us enough time to get upriver to Carl's."

She looks at me, then Carl.

"I don't think this is a very good idea," sings Carl. I understand his concern. Letting Carl's net soak for another hour of the flood could be a risk. When the fish are hitting this hard it is vital to work the gear, clear the fish. Otherwise the net becomes too heavy with fish.

"Joe!" barks Sharon. He is desperately trying to squeeze a fish out of the net.

"Yeah," he utters though gasping in his struggle.

"Use a pick. It will save your wrists," instructs Sharon.

Immediately after overhauling the nets onto Sharon's boat we then overhaul them again into the skiff with no motor, leaving 7,000 pounds of fish in Sharon's skiff.

Then we anchor Sharon's boat in the channel and jump into Carl's skiff, which has no nets or fish on board. There is about an hour and a half of good daylight left as we motor upriver to Carl's site. I hunker down in the bow and pull my hood over my head, cherishing the few moments of rest. My forearm tendons are a mess and the ibuprofens wore off long ago. I am quietly praying that the surge of fish never made it past all the other nets; between Carl's net and Sharon's are more than ten nets, surely enough to cork off the run. As the engine slows I stand up and look for the state of Carl's bobbing corks, always a way to see if the net is heavy with fish.

"They're gone," I say, pointing toward the net, but aware that the running line is still attached to the stakes onshore.

"Completely sunk," says Carl, referring to the fact that the net is so heavy with fish it has taken the corks underwater.

"We've got two hours to get this thing out of the water," says Sharon, grabbing the net and bringing the first few feet on board, flopping the first dozen fish onto the deck.

For the next hour we work nonstop, pulling inch by inch, closer and closer to the shore. As we pass the midway mark the wind dies down and immediately the mosquitoes descend upon us. Sharon reaches into her wet bag and pulls out four head nets, which we immediately pull over ourselves.

Pull. Pick. Pull. Pick. Every inch takes all the strength of four bodies to bring on board. The ebb tide is now ripping, making each pull that much harder. It doesn't take long before we are stuck in the place we stand, unable to free our legs from the thousands of pounds of fish wrapped around them like wet cement.

Darkness comes quickly and the shore is still twenty feet away. Now Sharon pulls out four headlamps, allowing us to keep working.

Finally we finish, a few minutes before closing. I take a knife from my waders and cut the rope holding the net to shore, freeing the boat. Sharon motors out to the channel and we pick up the net there—still attached to our anchor in the river—and overhaul it into the boat and quickly steer downriver. For once, on the day of the peak, all our gear is in the boats by closing.

Meanwhile a fog has settled on the river, reducing our visibility to less than fifty feet. We pass Dave's site but can't see Dave. When we finish, we need to check on him. A day like this can be overwhelming for a one-man operation.

Everything is going our way until we reach Sharon's site, where we left the skiff with no motor. It's gone. We motor around, visibility now down to twenty feet. To top it off, the temperature is dropping fast, turning our sweat cold against our bodies, creating the first stages of hypothermia.

After ten minutes we find the skiff, downstream. In almost complete darkness we quickly load Carl's net into the skiff, plus a few hundred pounds of fish from Carl's skiff. Then we tie a rope from the tender skiff to Carl's skiff, which by now is so heavy with fish it is barely riding above the waterline. Joe jumps on the tender skiff and readies himself to be towed. Carl slowly pulls away from our boat. I pull the anchor on Sharon's skiff and something strange happens: we don't move with the current. In a few seconds Sharon knows something is wrong.

"CARL," yells Sharon.

I turn around and see Sharon, her face near the engine.

"CARL," she yells again.

There is a faint sound of another approaching motor, but I can't see where it is coming from.

"We're sideways! We are sideways! Taking on water!" yells Sharon.

I look down and see water pouring in from the back of the boat. It is flowing forward, through the fish holds. Fast.

Going sideways in a river that is ebbing downstream can only mean one thing. Something has snagged the prop; most likely a net. We only have a few moments to get free before we will sink. I jump back and forth in the front of the boat, my mind racing with escape options if the boat swamps. I know going into the water is my last option. Nets surround us, and falling into one means certain death. The float coats—coast guard–mandated life jackets for commercial fishing boats—are buried under a pile of fish near the steering column. It would take five minutes to locate them.

"Get back here," yells Sharon at me. "I need some goddamn help."

I jump over the fish and into the rear of the boat in one leap. Now I hear both Carl's engine and a motor from a boat I can't see.

Finally Sharon rears up, her arms soaking wet.

"Carl!" she yells into the night, and then turns to me. "It's on the prop. Hold my feet."

I hold her while she punches her head and arms through the water to her shoulders. Carl has returned but is waiting for Sharon to reappear. The boat is only inches from sinking. Then Sharon snaps out of the water, and a few seconds later the boat flows with the current, downstream. We are free.

Just then a skiff appears out of the fog. It's the crew from the site below us. It is their net that was caught on our prop, and they have been following the net upstream through the fog looking for the snag. Everyone says a quick hello and in seconds they are gone, back into the fog and quickly overhauling their net.

As we cross the river in the dark I keep Carl in sight, just off to our right. Behind him the motorless skiff full of nets sways left to right, until Carl yells for Joe to move to the rear of the boat, helping to steady the weight.

"Sharon," I say.

"Yeah," she answers, exhausted and lighting a cigarette as she steers across the river.

"Didn't enjoy that."

"You think?" she replies, sighing. "Fuck, we hire someone for this day and it still doesn't matter."

Carl anchors the net skiff near where we load and unload every day, just below Sharon's cabin. Then Carl and Joe immediately head for the tender, to deliver the fish on Carl's skiff. It's two in the morning and hundreds of boats are racing around the river. The crab lights of the three tenders light up the night, revealing a small armada of boats and humans.

I take the lapse in action to quickly change into a dry sweatshirt.

And then Sharon reaches to start her engine. It won't turn over.

We figure water soaked the engine and fried the circuits. Quickly we pitch fish forward to free up space around the motor and give Sharon more room to work. Using her screwdriver, she attempts to jump the spark plugs. She tries for forty-five minutes straight. I can see Carl in line at the tender, under the glow of the massive crab lights, but there is no chance of his seeing or hearing us. We are lost from view in the pitch black water near the bank, and the cranes and engines at the tender are too loud for us to yell.

After resting for fifteen minutes, around 3:00 A.M., Sharon tries again and the engine kicks over. Sharon slams the engine in gear and in seconds we bolt for the tender. There we tie up to Carl, who happens to be next in line. The rules are not clear on waiting in line, but as long as the two boats are partners in paperwork, meaning they share the profits of the catch, they are usually allowed to cut in line. Behind Carl are a dozen set net skiffs. On the other side of the tender twenty drifters idle in the raging ebb water, waiting to deliver.

After we unload 15,000 pounds in thirty minutes, the captain of the tender yells to the people behind us, "We're full. Can't take any more."

THE BOATS STILL in line waiting to deliver have two choices. They can either head downriver and out to the bay, where dozens of tenders are waiting to take fish, or they can wait until the next day around noon, when the replacement tender rides into the river on the high tide. Either way they can't stay here tied up, for this boat will leave as soon as the high tide begins to flood. It will race to Naknek, where it will deliver the fish to the canneries.

Thankful we got rid of our fish, I grab a hose from the tender and wash down all the blood and slime on the skiff. After working for almost eighteen hours, after a twelve-hour day before that, and a fifteen-hour day before that, the mind and body tend to behave as if somewhat inebriated. It's five in the morning and the sun is rising over the tundra, but the exhaustion and inability to make clear decisions make me feel like I'm standing in a raft in the middle of the Pacific, lost at sea.

"Opening in one hour," says Carl, as we head toward shore. Sharon makes a few grunts and says nothing. Ignoring him is her way of challenging him to convince her that fishing back-to-back openings is what he really wants to do.

Before I throw anchor, we notice the skiff with the nets on board is sunk halfway up the railing. The nets are floating but still on board. There must be a hole in the boat, but we aren't going to locate it now. We spend the next twenty minutes towing the skiff high onto shore and overhauling the nets, for the eighth time today, into Carl's skiff.

We miss the following day's opener. Instead I wake around three in the afternoon. It is a sunny Alaskan day. Tonight is steak and shrimp dinner at the AGS cafeteria, and it makes me giddy to know that a good meal is in my future. Thirteen dollars, all you can eat. I will feast.

Outside, before dinner, I find Joe mending nets.

"I never thought I'd be here, in Alaska fishing," he says, happy in the moment.

"What do you think of it?"

"Last night was intense. I'm sore, but I liked it. Think I'll try cod on the Aleutians this fall. Maybe crabbing."

His face, although tired, beams with anticipation.

After dinner Carl and I go down to the river and patch the hole near the motor mount on the motorless skiff. Sharon is in the loft doing paperwork, probably realizing that even with the 15,000 pounds from last night this is going to be a tough year financially.

As Carl and I seal the hole in the skiff, we go over the events of last night. Taking water on board isn't something I want to ever live through again, and I wonder if this will finally be my last year.

Then I remember another scene from last night: Joe seated in the back of the skiff, probably ecstatic to have survived his first peak. He never knew he was sitting in a boat with a hole in it, crossing a river that has killed so many before.

I think, how long does one stay lucky in a place like Egegik?

PADDLING HOME

THE END ALWAYS BEGINS WITH A GRADUAL LOSS OF ENERGY TO do the work. Usually it happens within ten days after the peak. Getting out of bed becomes difficult; the body aches, begging for a week of long sleeps and hot baths. We never speak of it directly, or even realize we are about to quit. Around camp the talk is about how the fish have slowed down, and eventually the money spent catching the fish—gasoline, taxes, paying crew, et cetera—begins to equal the money earned in fish caught. Fishermen don't like spending money just to break even. After a few days of this, apathy sets in and people start to talk of what they are doing next. Crews begin fleeing the camp as fast as the small fleet of planes will take them.

And then, suddenly, one day it's just over.

It is mid-July, a week after the peak. The fleet has already gone from nine hundred boats to two hundred in the past week. In the nets we've been catching more flounder than sockeye, a clue the run is over. We catch floaters too, also known as ghost fish. These fish die in one net, drift out, and then spend several days floating back and forth with the current, getting tossed out of several nets a day. Their stench of rot and death is unmistakable, and when one is accidentally brought on board there is an urgent panic to get it off the skiff as soon as possible.

"I've seen people throw up when floaters are on board," says Sharon one day when the stench of the dead fish hovers over the river like the reek of an abattoir.

Tensions grow, and restless crews don't have enough work to keep them out of trouble. A few nights ago a fight broke out just behind the post office, in the Ivy camp. Carl and I are there drinking with several other set netters when two local men get into a fight with two young men from Ivy's crew. At first it is pure fists and bloody noses, but then others jump into the melee.

At some point, no one is quite sure when—due to drinking and the fact that no one carries a watch, phone, or much else other than a pocketknife—someone pulls out a gun and shots are fired.

An hour later, back at Sharon's, we see Jeremy, one of the Native men in the fight, walking up the road, back toward the Ivys'. He has a club in his hand and his friend, with the bloody nose, has a large stick. Carl walks out to counsel the angry men, both of whom he has known for years.

"I don't know, Jeremy. Maybe that's not such a good idea," Carl says, along with a laugh.

"Those motherfuckers got to pay."

"Okay. We'll be here."

The men disappear over the knoll, into the darkness. Even then we can hear them yelling into the night, haunting shrieks of injustice.

The next day the state troopers fly into camp and interview several people. They try to locate Jeremy. Too late, they are told. He has taken his drift boat down to Ugashik, forty miles down the coast, in the direction of Russia.

In the following days we skip the openings, too tired to care and doubtful whether the pounds caught will be worth the effort. Meanwhile, Dave Hart is catching decent loads each opening we sit out. As usual this becomes too much for Sharon. One afternoon she declares that we will fish the next opening.

So we set our alarms for 1:30 in the morning. The nets are already in the water, set out earlier that evening. We just have to pick the fish, preferably before high water.

I wake first. Sharon is asleep in her bed and Carl is downstairs sound asleep on the couch. I take a few minutes and look at them both, my mind racing through the four summers we have spent together. Flashes of anger, of panic, of laughter, of work all blur in front of my mind's eye.

Standing in the cold morning air, with my hands to my face trying to get my blood warm, I feel the deepest affection for their faults. For it is the sharing of our mistakes, fears, insecurities, and our hidden secrets that creates a sacred bond, the kind that is not broken easily. At this moment I realize I will know both of them the rest of my life.

I wake Sharon first, but of course she goes back to sleep immediately. Carl wakes and we make hot water and have bowls of cereal. Carl begins talking of not fishing next year. This is what he does every year. He also says he hopes to stop working at the lodge next summer. This desire for another line of work is not only about money. He wants to stop abusing his body.

He has a good plan. Instead of working for the lodge, he wants to fly people directly from King Salmon on floatplanes to the rapids by the lake. That is where the best fishing is, not to mention the beauty everyone seeks on their onetime visit to Alaska. Besides, the boat ride up the river is too hard on the clients, and it burns a lot of fuel. He has already secured a cabin as a place for his base of operations, only a few minutes from the rapids. I have no idea if his plan will work, but I do know that if Carl decides to go forward he will succeed because people enjoy being around him.

The coffee is percolating when Sharon stumbles down the stairs, her hair crazy from sleep and her eyes sagging under the weight of a season of work. Sharon waking always reminds me of a mother bear waking from a winter's sleep: groggy, cranky, and confused.

"Why are you always the last one out of bed?" I ask.

"Oh fuck off. You love it. It gives you time to get up and do all your little anal-compulsive routines, and then I get up just in time to drink some coffee and we are gone. It works out perfect for everyone," she says, with a large grin.

"The beautiful thing is that you actually believe that," I reply, laughing.

Carl laughs along.

"What? You don't want to go out?" shouts Sharon, at Carl.

"Not worth it," says Carl.

"Oh come on. Let's go pick the nets and come back and Bill will make us breakfast. He brought those goddamn tortillas all the way from Mexico or wherever."

Sharon looks at me. "What? You did, didn't you? Let's eat 'em."

"We're too late. Slack water was twenty minutes ago," says Carl, teasingly.

"Let's just quit then. Call it a season," says Sharon defiantly, pushing down violently on the coffee bean grinder.

Carl rolls his eyes and heads for the door.

"Five minutes," says Carl. "Five minutes or we're not going."

"Every time. All I have to do is threaten not to fish and he will fish," says Sharon, happy with herself.

In ten minutes we are dressed and down at the boat. We already missed high water, but we can still pick the fish before the tide turns. Then we will turn around and come home. We do this routine on most night openings—let the nets keep working in the flooding tide, while we return to the cabin and sleep in our own beds for a few hours.

It is pitch black. The moon glows behind a bank of clouds, and a fog hovers over the icy cold river.

"Remind anyone of anything?" I ask no one in particular.

"Oh, it's not the same," says Sharon.

• • •

LAST YEAR WE went out on a night just like this night, cold and foggy. That night we had debated for an hour whether we should go out or stay at home. We went out and spent three hours lost in the fog, finally returning to the cabin without putting one fathom of net in the water.

Learning from that foggy night, Carl and I load the boat with fog lights and extrawarm clothes. Sharon makes sure our float coats are on board.

As we slip into the fog I turn to Carl, who pulls out his expensive GPS, a tip from a happy hunter last season.

"What's wrong?" I ask, noticing he put away the GPS device.

"Battery's dead." He laughs under his breath, flashing me a giant smile. "I mean, come on. I don't need that piece of shit. You know that. Come on now. It's me."

I'm almost sure that if Carl were on an airplane crashing to earth he would find a way to convince all the screaming people on board not to worry. It is not that he thinks he is infallible. He knows he is not, but he also just doesn't see the point of worrying about worst-case scenarios.

As we approach the opposite bank I lean over the front of the boat to grab the net. Joe leans with me. Just then I see movement on the bank. Carl shines a light. Ten feet away, a grizzly sits on its hind end, staring at us. The bear makes a few heartless attempts to dodge the light and then runs off.

Over the years I have spent many hours walking the deep paths that line the riverbanks, etched into the earth by the daily stroll of the mighty grizzly as it saunters up and down the river looking for easy food. The paths don't change much from year to year and except for adjusting for the occasional slipping away of the bank, it can be said with great certainty that these trails have been here for thousands of years. Recently—in the last fifty years—the grizzlies have altered their scavenging rituals to accommodate the ever-present red ball: the fishing buoy. All up and down the river, buoys line the bank, indicating that an anchor is attached somewhere in the water. These buoys mark where fishermen begin each day of work. They are also beacons to the bears telling them a net has been at work in this particular spot and most likely dead fish will be swirling in eddies nearby. So, every morning—without fail—tracks go from the ancient trail on the grassy bank down through the mud to the river, and back. On a few occasions, usually in the morning, when the opening is at four or five, and the sun is still a distant blue, I have seen giant bears batting buoys or even attempting to sit on them. Playing.

Two hundred pounds of salmon are in Sharon's net, not enough to justify keeping it in the water. We turn the boat around, untie the net, and quickly overhaul it and grab both anchors. With that done we slowly make our way upriver to get Carl's net. In the past hour, the fog has begun to lift. A large half-moon has risen above the fog and throws a soft vanilla glow across the water. The water is calm tonight, almost glassy. We can see our breath, but I don't feel cold.

When we get to Carl's site—guided by the flashing light of his buoy—his net is gone. Upriver, Earl's net is floating in the correct spot, and below us Club Ohio's gear is in the water. With a floodlight we search the shore and then drive out into the river to the outer buoy. There it is: the net has broken free of the shore and now points directly downstream, still attached to the outer buoy.

"Bring it in, let's go," says Sharon, with the voice of a captain.

Carl drives and Sharon works the floodlight as Joe and I quickly overhaul the net into the boat. There are only a few fish, which we pick as the net comes in. It isn't hard work, but still I feel exhausted. The season has taken its toll.

"Okay, let's have some breakfast," says Carl, taking a puff of his cigarette.

We all sit and look across the river at the bank of lights that is Egegik.

"No fish, but hey, it's a beautiful night," says Carl with a laugh.

And then Carl turns over the engine. It won't start.

Ten minutes later Sharon and Carl stand over the engine with a flashlight, talking in soft voices as they try all the tricks they've learned over the years. Of course the screwdriver is applied, but even that doesn't work. We have enough gas. We have battery juice. We even have extra spark plugs, which they insert.

If this had been my first year I might be worried. Not now, not after all the time on the water with these two. I sit on a five-gallon bucket and let my mind drift, the moon gaining altitude like a slow-rising balloon.

After half an hour the ebb tide picks up more steam, and we are drifting downstream toward Egegik, the outboard motor acting as our only rudder.

"Well," says Sharon, from the back of the boat. "If we start paddling across the river, we should be able to make it over the sandbar before low water. Make it back by dawn."

"We? We as in Joe and I," I say, over my shoulder.

"Well, someone has to keep working on the engine," replies Sharon, making Carl laugh.

And so I grab the only paddle on board and begin to row. Every twenty strokes I hand it to Joe and he paddles. And every so often Carl tries the engine, but gets no response. Sharon figures it never recovered from the night we took on water and flooded the engine.

As we paddle across I turn my attention to the water.

Rivers have a way of seducing us into thinking they have been exactly the same since the beginning of time. The soft, constant flow of their water suggests eternity. We humans have a way of believing nature was created for us, and all we have to do is take from it, when really we are but gnats skimming along the skin of the river's daily flow.

Other than the very few mountain men determined to live off the land, no one travels to bush Alaska unless they are paid to come here—welfare, HUD housing and IRS agents, teachers, doctors, and police—or working to extract the resources from the land—

fishermen, researchers, oil wildcatters, or gold miners. We outsiders come north with an urgent need to help, arrest, or profit. This attitude of being the social and economical manager of nature and man alike comes with a price.

Maybe we do all this unknowingly, because we suffer from a secret wound, a pain that began long ago: when we decided to separate ourselves from nature. This wound is a secret, even to ourselves, and reveals itself only when one is faced with something so real, so removed from our modern world, that we can't deny it. The secret manifests in us like a surge of freedom; I feel it every time I step off the plane onto the gravel of Egegik. Or when I see a grizzly or spot wolf tracks.

Carl feels it. So much so that he decided to marry into it and live here full-time. Joe and Malibu Marty feel this freedom surging through their veins and are determined to find a way not to lose it. Sharon has made a life of coming back to it, year after year. Can I blame them? No, I only wish I could find it in more places in the world.

Someone once asked me to give him one word that describes fishing in Alaska. I replied, without too much thought: abundance. By this I meant that there are still places in the world where a person can sit on the bank of a river and throw a small net in the water and catch enough fish to feed four families, not once but several times over. This sounds like a myth, a fable from long ago, when the human and animal worlds were on more equal footing. But it does exist here in Egegik, for now.

We paddle until the sun creeps over the horizon, finally exposing us to other fishermen out to check their nets at low water. We are almost across when a fisherman pulls up beside us.

"Need a tow?" he asks, casually, and throws us a rope. I've seen the man around camp for four seasons but have never met him. He doesn't ask any more questions, just tows us to the shore. He isn't interested in the backstory of why we need a tow or why we have been paddling across the river all night. He has heard it before, or more than likely has had the same problem once. That's the thing about fishermen. We share a common story, a thread that runs through our lives. As we approach the bank I feel like I've fully arrived in this place. I relish the silence. I feel connected to these people, to this river. It has taken many years but no longer do I feel like an outsider or a greenhorn. I am part of those many threads that weave throughout the story of fishing in Alaska.

With this in mind, it seems fitting that in a time of jet boats, high technology, and speed-dial living we have paddled across the river, slowly but with confidence, not unlike something a Native fisherman would have done two thousand years ago.

Once we tie off the boat we grab our belongings and stand onshore, shaking off the cold.

"Well?" asks Sharon.

"Guess so," I say, and look at Carl.

"Okay," says Carl, shaking his head.

Confused, Joe is loaded with more gear than the rest of us. He drags behind as Sharon, Carl, and I walk the river's edge, not saying anything for a few minutes.

"I'll call you in the spring," says Sharon, staring straight ahead.

We laugh together, and then go silent and keep walking. Finally we turn up the bank, toward the cabin. Still walking, I look over my shoulder one last time at the Egegik River. As always, it keeps rushing out to the sea.

And just like that, the season ends.

ACKNOWLEDGMENTS

This book is written with the greatest respect and appreciation for the people of Egegik, including the ones that I never met and the ones who didn't care if I was ever there.

In describing the people from Egegik I often used the word *Native,* a word sometimes loaded with pejorative connotations. I considered using other words, but Native is the word most often used by the people of Egegik to describe themselves. And yet, some may take offense at my use of the word. I can only refer to the *Webster's International Dictionary* as to my intent:

> Native: Belonging to one by nature; belonging to or associated with a particular place by birth.

I want to thank Cindy Rhodes for introducing me to Egegik, and the Hart family for accepting me into their camp: Warren, Dave, Ron, and Lynnie. In Egegik many thanks to the Deigh family: Jannelle, Dick, Scovi, Mike, and Kevin. They provided insight, laughter, and tough love. Thanks to Bunny Alto at the post office. In King Salmon, my thanks to Slim Horsted at the Department of Fish and Game for his expertise.

Thanks to the Mesa Refuge for the sanctuary; as always, it was productive. Thanks to Rod Kass and Sally Holcomb for giving me shelter in Mexico to work on the book. I want to thank my agent, Betsy Lerner, for her persistence in getting this story into print. At Scribner, thanks to Karen Thompson for keeping it all on track, and to my editor, Colin Harrison, who edited with the greatest care and respect for the work and people in the story. Finally I wish to say thanks to all the editors who checked and rechecked the facts of the book; any remaining errors are entirely my own.

My deepest thanks and respect extend to Sharon Hart and Carl Adams for sharing their knowledge and skills, but most of all for their friendship. They opened up their world to me and this book is a testimony of a trust formed through the acts of hard work, exhaustion, and laughter.

Finally, I want to thank my wife, Leigh. Without her this book would not exist.

ABOUT THE AUTHOR

BILL CARTER is the author of *Boom, Bust, Boom: A Story about Copper, the Metal That Runs the World* and *Fools Rush In: A True Story of Love, War, and Redemption*. He is also the director of *Miss Sarajevo*, an award-winning documentary produced by Bono. He has written for *Rolling Stone*, *Outside*, *Men's Journal*, and other publications. He is an associate professor of practice at Northern Arizona University and lives with his family in Sedona, Arizona.